Math Contests for High School
Volume 7

School Years: 2011-2012 through 2015-2016

Written by

Steven R. Conrad • Daniel Flegler • Adam Raichel

Published by MATH LEAGUE PRESS
Printed in the United States of America

Cover art by Bob DeRosa

First Printing, 2016
Copyright © 2016
by Mathematics Leagues Inc.
All Rights Reserved

Math League Press
P.O. Box 17
Tenafly, NJ 07670-0017

ISBN 978-0-940805-23-1

Preface

Math Contests—High School, Volume 7 is the seventh volume in our series of problem books for high school students. The first six volumes contain contests given in the school years 1977-1978 through 2010-2011. Volume 7 contains contests given from 2011-2012 through 2015-2016. (Use the order form on page 70 to order any of our 21 books.)

These books give classes, clubs, teams, and individuals diversified collections of high school math problems. All of these contests were used in regional interscholastic competition throughout the United States and Canada. Each contest was taken by about 80 000 students. In the contest section, each page contains a complete contest that can be worked during a 30-minute period. The convenient format makes this book easy to use in a class, a math club, or for just plain fun. In addition, detailed solutions for each contest also appear on a single page.

Every contest has questions from different areas of mathematics. The goal is to encourage interest in mathematics through solving *worthwhile* problems. Many students first develop an interest in mathematics through problem-solving activities such as these contests. On each contest, the last two questions are generally more difficult than the first four. The final question on each contest is intended to challenge the very best mathematics students. The problems require no knowledge beyond secondary school mathematics. No knowledge of calculus is required to solve any of these problems. From two to four questions on each contest are accessible to students with only a knowledge of elementary algebra. Starting with the 1992-93 school year, students have been permitted to use any calculator without a QWERTY keyboard on any of our contests.

This book is divided into four sections for ease of use by both students and teachers. The first section of the book contains the contests. Each contest contains six questions that can be worked in a 30-minute period. The second section of the book contains detailed solutions to all the contests. Often, several solutions are given for a problem. Where appropriate, notes about interesting aspects of a problem are mentioned on the solutions page. The third section of the book consists of a listing of the answers to each contest question. The last section of the book contains the difficulty rating percentages for each question. These percentages (based on actual student performance on these contests) determine the relative difficulty of each question.

You may prefer to consult the answer section rather than the solution section when first reviewing a contest. The authors believe that reworking a problem, knowing the answer (but *not* the solution), often helps to better understand problem-solving techniques.

Revisions have been made to the wording of some problems for the sake of clarity and correctness. The authors welcome comments you may have about either the questions or the solutions. Though we believe there are no errors in this book, each of us agrees to blame the others should any errors be found!

Steven R. Conrad, Daniel Flegler, & Adam Raichel, contest authors

Acknowledgments

For the beauty, cleverness, and breadth of his numerous mathematical contributions for the past 30 years, we are indebted to Michael Selby.

For her continued patience and understanding, special thanks to Marina Conrad, whose only mathematical skill, an important one, is the ability to count the ways.

For demonstrating the meaning of selflessness on a daily basis, special thanks to Grace Flegler.

To Daniel Will-Harris, whose skill in graphic design is exceeded only by his skill in writing *really* funny computer books, thanks for help when we needed it most: the year we first began to typeset these contests on a computer.

Table Of Contents

The Contests

October, 2011 – March, 2016

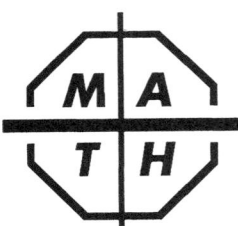

HIGH SCHOOL MATHEMATICS CONTESTS

Math League Press, P.O. Box 17, Tenafly, New Jersey 07670-0017

Contest Number 1 *Any calculator without a QWERTY keyboard is allowed.* Answers must be exact *or* have 4 (or more) significant digits, correctly rounded. **October 18, 2011**

Name _____ Teacher _____ Grade Level _____ Score _____

Time Limit: 30 minutes   *Answer Column*

1-1. What is the only ordered pair of positive integers (m,n) for which $$\sqrt{m} + \sqrt{n} = \sqrt{4} ?$$ *(handwritten: 1 under m, 9 under n)*	1-1. $(1, 9)$
1-2. How many values of x satisfy $(x-2010)(x-2011) = (x-2011)(x-2012)$?	1-2.
1-3. All 26 letters of the English alphabet were evenly spaced alphabetically on a circular road. A rabbit hopped on the circular road at A and continued hopping, skipping 2 letters at a time, so that the next 3 letters to which it hopped were D, G, and J. If the rabbit stopped hopping the first time it had hopped to every letter once, then at what letter did the rabbit stop hopping?	1-3. $\times$
1-4. In the largest right triangle shown, the altitude to the hypotenuse has a length of 12. If every line segment in the diagram has an integral length, what is the perimeter of the largest right triangle? *(diagram of right triangle with altitude labeled 12)*	1-4.
1-5. At work I earned $\$X$ on my 1st day and $\$Y$ on my 2nd day (where X and Y are integers). After that, on Day N, I earned the sum of my earnings on Days $N-1$ and $N-2$. If I earned $\$55$ on my 10th day at work, then how many dollars did I earn on my 3rd day at work?	1-5.
1-6. What is the *exact* value of the smallest integer x greater than 1 million for which the sum of 2^x and 3^x is divisible by 7?	1-6.

HIGH SCHOOL MATHEMATICS CONTESTS

Math League Press, P.O. Box 17, Tenafly, New Jersey 07670-0017

Contest Number 2 *Any calculator without a QWERTY keyboard is allowed.* Answers must be exact *or* have 4 (or more) significant digits, correctly rounded. **November 15, 2011**

Name _____ Teacher _____ Grade Level _____ Score _____

Time Limit: 30 minutes   *Answer Column*

2-1.   What is the largest possible value of $(x^2 - 2x - 1)^3 + (1 + 2x - x^2)^3$? | 2-1.

2-2.   Two vertices of a square are at (3,0) and (5,0). The other vertices lie in the first quadrant. What is the slope of the line through the origin that splits the square into two regions of equal area? | 2-2.

2-3.   So far this year, the number of kilometers I have flown equals the sum of two consecutive integers whose squares differ by 2011. How many kilometers have I flown so far this year? | 2-3.

2-4.   The "double Y," whose 6 angles are all congruent, and whose 4 non-vertical segments are congruent, is the shortest network that connects all 4 vertices of a square. If the length of each side of a square is 2, what is the total length of its "double Y" network? | 2-4.

2-5. N is a 5-digit whole number. The 6-digit number $N1$, formed by placing the digit 1 after N, is 3 times as large as the 6-digit number $1N$, formed by placing the 1 before N. What is the 5-digit number N? | 2-5.

2-6.   The number 1 can be written as a sum of n positive numbers (not necessarily distinct), each of which has a decimal representation that consists entirely of 8s, or entirely of 0s and 8s. What is the least possible value of n? | 2-6.

Solutions on Page 35 • Answers on Page 66

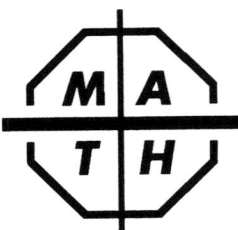
Contest Number 3 *Any calculator without a QWERTY keyboard is allowed.* Answers must be exact *or* have 4 (or more) significant digits, correctly rounded. **December 13, 2011**

Name _____ **Teacher** _____ **Grade Level** ____ **Score** ____

Time Limit: 30 minutes   *Answer Column*

3-1.	What is the greatest difference in the perimeters of two rectangles whose sides have integral lengths and whose areas are both 25?	3-1.
3-2.	What are five positive integers whose mean and median are both 5 and whose only mode is 8?	3-2.
3-3.	If x and y are acute angles whose sum is 60°, what is the largest possible value of $(\tan x)(\tan y)$?	3-3.
3-4.	In a 75 m race, if Ace gives Flash a 15 m head start, Ace and Flash will tie. If Flash gives Speedy a 15 m head start, Flash and Speedy will tie. At these rates, how many meters head start should Ace give Speedy for Ace and Speedy to tie in a 75 m race?	3-4.
3-5.	If f and g are linear functions such that $g(f(x)) = 2x+6$, and the graph of $y = f(g(x))$ passes through the origin, what is the value of $f(g(2011))$?	3-5.
3-6.	Two perpendicular diameters are drawn in a circle, as shown. From an endpoint of one of them, chords that intersect the other diameter at different points are drawn as shown. The horizontal diameter splits one of the chords into a 2:1 ratio and the other into a 3:1 ratio. What is the ratio, bigger to smaller, of the lengths of the two chords?	3-6.

Solutions on Page 36 • Answers on Page 66

HIGH SCHOOL MATHEMATICS CONTESTS

Math League Press, P.O. Box 17, Tenafly, New Jersey 07670-0017

Contest Number 4 *Any calculator without a QWERTY keyboard is allowed.* Answers must be exact *or* have 4 (or more) significant digits, correctly rounded. **January 10, 2012**

Name _____ Teacher _____ Grade Level _____ Score _____

Time Limit: 30 minutes

Answer Column

4-1. What is the average of 2012 consecutive positive integers whose sum is 2012^3?

4-1.

4-2. When n is a positive integer, $\frac{1}{n}$ and $\frac{1}{n+1}$ are called a pair of *consecutive unit fractions.* ($\frac{1}{7}$ and $\frac{1}{8}$ are such a pair.) What are two consecutive unit fractions respectively greater than and less than $\frac{3}{89}$?

4-2.

4-3. Anytime each of three consecutive months has exactly four Fridays, my birthday will fall in one of those three months. Which month is that?

4-3.

4-4. In the diagram shown at the right, six squares are arranged into a 3×2 rectangle. Line segments $\overline{AB}$ and $\overline{BC}$ are drawn. What is $m\angle ABC$?

4-4.

4-5. There are an infinite number of polynomials P for which $P(x+5) - P(x) = 2$ for all x. What is the least possible value of $P(4) - P(2)$?

4-5.

4-6. What are all ordered pairs of real numbers (x,y) that satisfy both $x^2-4xy+5y^2+2x-12y+17 = 0$ and $x^2-8xy+9y^2+18x-16y-31 = 0$?

4-6.

Solutions on Page 37 • Answers on Page 66

HIGH SCHOOL MATHEMATICS CONTESTS

Math League Press, P.O. Box 17, Tenafly, New Jersey 07670-0017

Contest Number 5 *Any calculator without a QWERTY keyboard is allowed.* Answers must be exact *or* have 4 (or more) significant digits, correctly rounded. **February 14, 2012**

Name _____ Teacher _____ Grade Level _____ Score _____

Time Limit: 30 minutes | *Answer Column*

5-1. If the square root of circle C's area is 2012π, how long is C's radius? | 5-1.

5-2. For what value of k are the roots of $x^2 + kx + 600 = 0$ consecutive negative integers? | 5-2.

5-3. In the sequence of non-zero numbers $x^1, x^2, x^3, x^4, \ldots$, the nth term is x^n. If, from the third term on, each term is the sum of the preceding two terms, what are all possible values of x? | 5-3.

5-4. I bought two lovebirds for different prices. I later sold them at a loss, for $24 each. My percent loss on each lovebird was numerically equal to its dollar price. What were the two *different* prices I originally paid for the lovebirds, in dollars? | 5-4.

5-5. In the diagram, 15 congruent squares are assembled as shown. A large triangle whose vertices are also vertices of 3 of these squares is partitioned into 3 smaller right triangles, as shown. For what ordered triple of numbers (a,b,c), with $0 < a < b < c < 10$, is this diagram a proof of the identity $\operatorname{Arctan} a + \operatorname{Arctan} b + \operatorname{Arctan} c = \pi$? | 5-5.

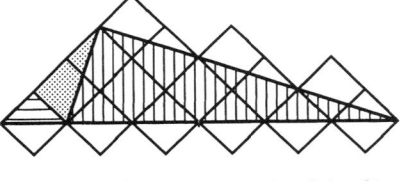

5-6. If a is real, what is the only real number that could be a multiple root of $x^3 + ax + 1 = 0$? | 5-6.

Solutions on Page 38 • Answers on Page 66

HIGH SCHOOL MATHEMATICS CONTESTS

Math League Press, P.O. Box 17, Tenafly, New Jersey 07670-0017

Contest Number 6   *Any calculator without a QWERTY keyboard is allowed.* Answers must be exact *or* have 4 (or more) significant digits, correctly rounded.   **March 13, 2012**

Name _____ Teacher _____ Grade Level ____ Score ____

Time Limit: 30 minutes   *Answer Column*

6-1. A certain line segment through the origin contains exactly two points whose coordinates are pairs of consecutive integers. One of these points is $(-2011, -2012)$. What are the coordinates of the other?

6-1.

6-2. Dad and his two children were all born on July 28, but in different years. On their most recent birthday, the sum of their ages was 40, and the age of each was a prime. How old was Dad that day?

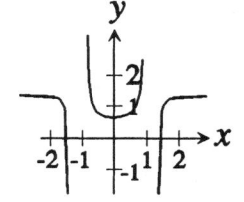

6-2.

6-3. What ordered triple of integers (a,b,c), with $c > b > a > 0$, satisfies
$$(\sqrt{a} + \sqrt{b} + \sqrt{c})^2 = 10 + \sqrt{24} + \sqrt{40} + \sqrt{60}?$$

6-3.

6-4. A portion of the graph of $y = f(x)$ is shown at the right. In the interval depicted, how many solutions are there to the equation $f(x) = \frac{1}{f(x)}$?

6-4.

6-5. If $x > -1$, then exactly 2 values of x make $\log_2\left((x+1)^{\log_2(x+1)^8}\right) = 8$. One of these is $x = 1$. What is the other?

6-5.

6-6. Balls are randomly removed from a bag without replacement. If the probability that the first 5 balls withdrawn are all green is one-half, what is the least possible number of balls in the bag at the start?

6-6.

Solutions on Page 39 • Answers on Page 66

Contest Number 1 *Any calculator without a QWERTY keyboard is allowed.* Answers must be exact *or* have 4 (or more) significant digits, correctly rounded. **October 16, 2012**

Name _____ Teacher _____ Grade Level ____ Score ____

Time Limit: 30 minutes | *Answer Column*

1-1. Two sides of a square are radii of the same circle, as shown in the diagram at the right. What is the ratio of the circumference of the circle to the perimeter of the square?

1-1. ~~4.13.04~~ $\frac{\pi}{2}$

1-2. If $|a-b| = 1005$ and $|b-c| = 1007$, what are both possible values of $|a-c|$?

1-2.

1-3. In a convex polygon with perimeter 100, if each side has an integral length, what is the maximum possible length of one side?

1-3.

1-4. What is the least positive integer that *cannot* be written as the difference between two positive primes?

1-4.

1-5. Reading from left to right, for how many integers greater than 9 is it true that every digit after the first exceeds the digit it follows?

[**NOTE:** As an example, one such integer is 24789.]

1-5.

1-6. One holiday, I gave each of my 3 grandsons x coins and each of my 4 granddaughters y coins. The total number of coins that I gave to my grandchildren will allow for only one ordered pair of positive integers (x,y). At most how many coins did I give to my 7 grandchildren?

1-6.

© 2012 by Mathematics Leagues Inc.

HIGH SCHOOL MATHEMATICS CONTESTS

Math League Press, P.O. Box 17, Tenafly, New Jersey 07670-0017

Contest Number 2 *Any calculator without a QWERTY keyboard is allowed.* Answers must be exact *or* have 4 (or more) significant digits, correctly rounded. **November 13, 2012**

Name _____ Teacher _____ Grade Level _____ Score _____

Time Limit: 30 minutes   *Answer Column*

2-1. What is the largest prime divisor of every 3-digit number with 3 identical non-zero digits?	2-1.
2-2. If 3 adult bears ate an average of 16 hot dogs each, and 2 bear cubs ate an average of 6 hot dogs each, then (for these 5 bears) what was the average number of hot dogs eaten per bear?	2-2.
2-3. A semicircle is tangent to both legs of a right triangle and has its center on the hypotenuse. The hypotenuse is partitioned into 4 segments, with lengths 3, 12, 12, and x, as shown. What is the value of x?	2-3.
2-4. What are all 3 ordered triples of integers (a,b,c), with $0 < a \le b \le c$, for which $\frac{1}{a} + \frac{1}{b} + \frac{1}{c} = 1$?	2-4.
2-5. A distribution consists of the integers from 1 through 100, inclusive, such that the frequency of each integer n is 2^{n-1}. What is the median of this distribution?	2-5.
2-6. If S is a 2012×2012 square split into unit squares, a diagonal of S will pass through the interior of 2012 unit squares. If R is a 2012×2015 rectangle split into unit squares, a diagonal of R will pass through the interior of how many unit squares?	2-6.

HIGH SCHOOL MATHEMATICS CONTESTS

Math League Press, P.O. Box 17, Tenafly, New Jersey 07670-0017

Contest Number 3 *Any calculator without a QWERTY keyboard is allowed.* Answers must be exact *or* have 4 (or more) significant digits, correctly rounded. **December 11, 2012**

Name _____ Teacher _____ Grade Level _____ Score _____

Time Limit: 30 minutes | *Answer Column*

3-1. The vertices of a rhombus are four points on two concentric circles, as shown. The lengths of the diameters of the two circles are 6 and 8. What is the perimeter of this rhombus?

3-1.

3-2. In a certain 2012-term sequence, the sum of any two consecutive terms equals the product of the same two terms. If the 2012th term is 2, what is the sum of all 2012 terms?

3-2.

3-3. For what pair of positive integers (x,y) will $(x-y-1)^2+(x+y-39)^2 = 0$?

3-3.

3-4. Five positive integers are written in increasing order, and the difference between adjacent terms is constant. If the sum of the integers is 540, what is the maximum possible value of the largest integer?

3-4.

3-5. In the coordinate plane, the vertices of a quadrilateral are $(-4,-2)$, $(8,5)$, $(6,8)$, and $(-1,6)$. What are the coordinates of the point in the plane for which the sum of the distances from this point to the vertices of the quadrilateral is a minimum?

C(6,8)
D(-1,6)
B(8,5)
A(-4,-2)

3-5.

3-6. A standard cubical die is thrown four times, with respective outcomes a, b, c, and d. What is the probability that $a \le b \le c \le d$?

3-6.

© 2012 by Mathematics Leagues Inc.

Solutions on Page 42 • Answers on Page 66

Contest Number 4   *Any calculator without a QWERTY keyboard is allowed.* Answers must be exact *or* have 4 (or more) significant digits, correctly rounded.   **January 8, 2013**

Name _____ Teacher _____ Grade Level _____ Score _____

Time Limit: 30 minutes   *Answer Column*

4-1.	A line crosses the y-axis at $(a+2013, a+2012)$. What is the value of a?	4-1.
4-2.	Diagonals are drawn in a 12×16 rectangle. As shown, a smaller rectangle has its vertices on these diagonals so that the distance from each vertex of the smaller rectangle to the nearest vertex of the larger rectangle is 5. What is the area of the smaller rectangle?	4-2.
4-3.	Two days ago, x dogs skated. Yesterday, exactly 20% more than x dogs skated. Exactly 40% more dogs skated today than skated yesterday. What is the minimum positive integer value of x?	4-3.
4-4.	For how many integers x, with $0 < x < 100$, is $\log_8 x$ a rational number?	4-4.
4-5.	A circle is circumscribed about an isosceles trapezoid whose bases have lengths 10 and 18 and are 4 units apart. The circle's center doesn't lie in the interior of the trapezoid. What is the circle's area?	4-5.
4-6.	Let $P(x) = x^5 + ax^4 + bx^3 + cx^2 + dx + e$, and suppose that $P(1) = P(2) = P(3) = P(4) = P(5) = 0$. What is the value of $1 - a + b - c + d - e$?	4-6.

Solutions on Page 43 • Answers on Page 66

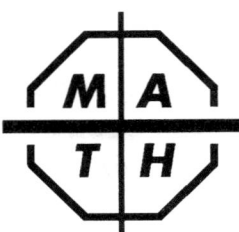
Contest Number 5   *Any calculator without a QWERTY keyboard is allowed.* Answers must be exact *or* have 4 (or more) significant digits, correctly rounded.   **February 12, 2013**

Name _____ Teacher _____ Grade Level _____ Score _____

Time Limit: 30 minutes | *Answer Column*

5-1. A regular hexagon, drawn as shown, has one vertex on the y-axis and one side on the x-axis. If the x-coordinate of the hexagon's rightmost vertex is 4, what is the hexagon's perimeter?

5-1.

5-2. What is the only integer x for which $x^2 - 100x + 2500$ is *not* positive?

5-2.

5-3. If I double 18, I get 36, the square of an integer. What is the largest integer less than 1000 whose double is the square of an integer?

5-3.

5-4. If the value of the infinite series $\cos^2 x + \cos^4 x + \cos^6 x + \ldots + \cos^{2n}x + \ldots$ is 2013, what is the value of the infinite series $\sin^2 x + \sin^4 x + \sin^6 x + \ldots + \sin^{2n}x + \ldots$?

5-4.

5-5. A bowl contains 100 cookies, each red or blue. Al chooses 2 cookies. If there are as many ways for Al to choose 2 cookies of the same color as there are for him to choose 1 of each color, what is the least possible number of red cookies in the bowl? [NOTE: Disregard the order in which the 2 cookies are chosen.]

5-5.

5-6. For every integer $N > 0$, there's at least one pair of integers (a,b), with $0 < a \le b$, for which $N = ab - (a+b)$. For example, $20 = 4 \times 8 - (4+8)$ or $20 = 2 \times 22 - (2+22)$. What are all positive integers greater than 40 and less than 50 that have only one such representation?

5-6.

HIGH SCHOOL MATHEMATICS CONTESTS

Math League Press, P.O. Box 17, Tenafly, New Jersey 07670-0017

Contest Number 6   *Any calculator without a QWERTY keyboard is allowed.* Answers must be exact *or* have 4 (or more) significant digits, correctly rounded.   **March 12, 2013**

Name _____ Teacher _____ Grade Level _____ Score _____

Time Limit: 30 minutes   *Answer Column*

6-1. The area of square A is 2013 more than the area of square B. The length of a side of square A is 1 more than the length of a side of square B. What is the perimeter of square A?

6-1.

6-2. After I factored 94 in 2 different ways, I used my results to find the only 2 pairs of positive integers (x,y) that satisfy $(x-y)(x+y-1) = 94$. One of these pairs is $(48,47)$. What is the other pair?

6-2.

6-3. Each of 8 players played a game of chess against each of the other 7. When the players tied, each got 0.5. If not, the winner got 1 and the loser 0. At the end of the 7 rounds, Pat's total was higher than any other player's. What is the least that Pat's total could have been?

6-3.

6-4. Points A and B are on line ℓ. A second line is parallel to ℓ. If the distance from A to B exceeds the distance between the lines, *at most* how many different points C can be selected on the second line so that $\triangle ABC$ is isoscles?

6-4.

6-5. There are only two pairs of positive integers (x,y) for which both $\frac{21}{x}$ and $\frac{70}{y}$ are in lowest terms and for which $\frac{21}{x} + \frac{70}{y}$ is an integer. One such pair is $(1,1)$. What is the other such pair?

6-5.

6-6. What is the area of a circle if two of its perpendicular chords divide each other into four segments whose lengths are 4, 6, 12, and 18?

6-6.

Solutions on Page 45 • Answers on Page 66   13

HIGH SCHOOL MATHEMATICS CONTESTS

Math League Press, P.O. Box 17, Tenafly, New Jersey 07670-0017

Contest Number 1   *Any calculator without a QWERTY keyboard is allowed.* Answers must be exact *or* have 4 (or more) significant digits, correctly rounded.   **October 15, 2013**

Name _____ Teacher _____ Grade Level _____ Score _____

Time Limit: 30 minutes | *Answer Column*

1-1. Two congruent squares overlap to form 3 congruent, non-overlapping rectangles, as shown. If the perimeter of each of these rectangles is 18, what is the area of each? | 1-1.

1-2. What is the greatest possible sum of two multiples of 12, each less than 100, whose greatest common factor is 24? | 1-2.

1-3. The right side of the equation $3(ABC) = BBB$ represents a 3-digit number with 3 identical digits. If different letters represent different digits, what is the ordered triple of non-zero digits (A,B,C)? | 1-3.

1-4. If $N = 10^{2013} - 2013$ is expanded and written as an integer in standard form, what is the sum of the digits of N? | 1-4.

1-5. What is the length of the hypotenuse of a right triangle in which the medians to the legs have lengths 3 and 4? | 1-5.

1-6. The Surfboard Store sells Special Surfboards that are made so carefully that only 1 in 1 thousand is bad. The store tests all Special Surfboards using a test that is 99% accurate. If my Special Surfboard tests bad, then, to the nearest 1%, what is the probability that my Special Surfboard is bad? | 1-6.

14 | Solutions on Page 46 • Answers on Page 66

Contest Number 2 *Any calculator without a QWERTY keyboard is allowed.* Answers must be exact *or* have 4 (or more) significant digits, correctly rounded. **November 12, 2013**

Name _____ Teacher _____ Grade Level _____ Score _____

Time Limit: 30 minutes | *Answer Column*

2-1. When written as 02-03-04, Feb. 3, 2004 consists of three consecutive integers whose sum is a perfect square. Writing your answer as MM-DD-YY, what is the first date after 02-03-04 for which MM, DD, and YY are consecutive integers whose sum is a perfect square? | 2-1.

2-2. If $a \neq b$, but $a^2 + a = b^2 + b$, what is the value of $a + b$? | 2-2.

2-3. What is the only odd prime factor of $2^{67} + 2^{71}$? | 2-3.

2-4. When I remove a corner from a cube of edge-length 6, the solid formed has three vertices that were not vertices of the cube. The distances from each of these vertices to the nearest remaining vertices of the cube are 1, 2, and 3, as shown. What is the volume of the solid remaining after removal of the corner? | 2-4.

2-5. I wrote a list of 100 positive integers whose sum and product are equal. Of the integers on my list, at most how many can be a 1? | 2-5.

2-6. In a certain quadrilateral, the three shortest sides are congruent, and both diagonals are as long as the longest side. What is the degree-measure of the largest angle of this quadrilateral? | 2-6.

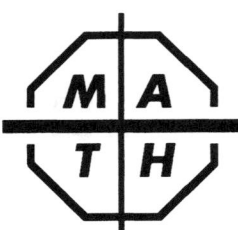

HIGH SCHOOL MATHEMATICS CONTESTS

Math League Press, P.O. Box 17, Tenafly, New Jersey 07670-0017

Contest Number 3 *Any calculator without a QWERTY keyboard is allowed.* Answers must be exact *or* have 4 (or more) significant digits, correctly rounded. **December 3, 2013**

Name _____ Teacher _____ Grade Level _____ Score _____

Time Limit: 30 minutes

Answer Column

3-1. One vertex of an equilateral triangle is at the center of a circle, and the other two vertices lie on the circle, as shown. If the circle's area is 16π, what is the triangle's perimeter?

3-1. ____

3-2. My uncle's birthday is today. He is older than I am, and he is less than 100 years old. If his age is six times the sum of its digits, how old is he?

3-2. ____

3-3. The first 4 rows of an array of consecutive integers is shown at the right. The first row has 1 entry. Every other row has 2 more entries than the row directly above it. What is the value of the 2013th entry in the 2013th row? [NOTE: For this question, provide an exact answer, not an approximate answer.]

```
        1
      2  3  4
    5  6  7  8  9
 10 11 12 13 14 15 16
```

3-3. ____

3-4. If the third-degree polynomial equation $P(x) = 0$ has three unequal real roots, what is the least possible number of unequal real roots there could be for the sixth-degree polynomial equation $P(x^2) = 0$?

3-4. ____

3-5. In an *arithmetic sequence* (such as 7, 12, 17, 22, . . .), the difference between successive terms is fixed. If the sum of the 72nd and 112th terms of one such sequence is 22, what is the sum of the first 183 terms?

3-5. ____

3-6. After five integers are paired in all possible ways, the integers in each pair are added. The ten sums obtained (not all different) are 1, 4, 5, 7, 8, 8, 11, 11, 14, and 15. What are these five integers?

3-6. ____

© 2014 by Mathematics Leagues Inc.

Solutions on Page 48 • Answers on Page 66

HIGH SCHOOL MATHEMATICS CONTESTS

Math League Press, P.O. Box 17, Tenafly, New Jersey 07670-0017

Contest Number 4   *Any calculator without a QWERTY keyboard is allowed.* Answers must be exact *or* have 4 (or more) significant digits, correctly rounded.   **January 14, 2014**

Name _____ Teacher _____ Grade Level _____ Score _____

Time Limit: 30 minutes   *Answer Column*

4-1. If each of the numbers 2, 3, 4, 6, 12, 18, 24, 36, and 72, is paired with a different one of the numbers 5, 10, 15, 20, 30, 60, 90, 120, and 180 so that the product stays the same, with which number is 30 paired?

4-1.

4-2. For some integers a and b, the same integer x satisfies both of the equations $x^2 - ax + 2014 = 0$ and $x^2 - bx + 2015 = 0$. What is the greatest possible value of this common solution?

4-2.

4-3. In a 10-km race, First Runner beat Second Runner by 2 km, and First Runner beat Third Runner by 4 km. If all three runners always ran at constant rates, by how many km did Second Runner beat Third Runner?

4-3.

4-4. The circle circumscribed about acute $\triangle T$ has area π. If the length of the longest side of $\triangle T$ is x, what is the least possible value of x?

4-4.

4-5. What is the least possible value of the smallest of 99 consecutive positive integers whose sum is a perfect cube?

4-5.

4-6. In the grid shown at the right, a group of 9 squares has been shaded. There's only one line through the origin that divides this shaded region into two regions of equal area. What is the slope of this line?

4-6.

Solutions on Page 49 • Answers on Page 66

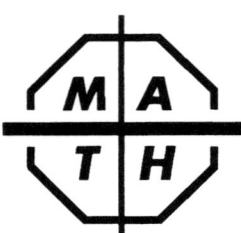

HIGH SCHOOL MATHEMATICS CONTESTS

Math League Press, P.O. Box 17, Tenafly, New Jersey 07670-0017

Contest Number 5 *Any calculator without a QWERTY keyboard is allowed.* Answers must be exact *or* have 4 (or more) significant digits, correctly rounded. **February 11, 2014**

Name _____ Teacher _____ Grade Level _____ Score ____

Time Limit: 30 minutes | *Answer Column*

5-1. What value of x satisifies $x^2-3x+2 = 0$ but **not** $x^2-x-2 = 0$? | 5-1.

5-2. A circle and a square share a common center, as shown. If the area of each is 2014, what is the difference between the area of that part of the circle that's outside the square and the area of that part of the square that's outside the circle? | 5-2.

5-3. If m is the maximum number of acute interior angles a convex quadrilateral can have, and M is the maximum number of obtuse interior angles a convex quadrilateral can have, what is the value of $m+M$? | 5-3.

5-4. The garbage dump manager has clothes pins in only 3 colors, and 1 more than half his clothes pins are red. If the red ones are removed, then 1 less than half the remaining clothes pins are green. If the green ones are also removed, then 8 more than half the remaining clothes pins are blue. How many clothes pins does the garbage dump manager have all together? | 5-4.

5-5. A certain polynomial P has the property that $P(z) = P(iz)$ for all numbers z, real or imaginary, where $i^2 = -1$. If the number 2 is known to be a root of $P(z) = 0$, what are all numbers, real or imaginary (including 2), which must be roots of $P(z) = 0$? | 5-5.

5-6. If a and x are real numbers, then, explicitly in terms of a, what real number x satisfies $\sqrt[3]{x+\sqrt{x^2+a^3}} + \sqrt[3]{x-\sqrt{x^2+a^3}} = a$? | 5-6.

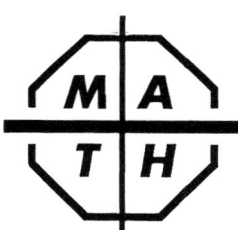

HIGH SCHOOL MATHEMATICS CONTESTS

Math League Press, P.O. Box 17, Tenafly, New Jersey 07670-0017

Contest Number 6 *Any calculator without a QWERTY keyboard is allowed.* Answers must be exact *or* have 4 (or more) significant digits, correctly rounded. **March 11, 2014**

Name _____ Teacher _____ Grade Level _____ Score _____

Time Limit: 30 minutes | *Answer Column*

6-1. What are both values of x that satisfy $x^2 - x = 2^2 - 2$? | 6-1.

6-2. Circles of radii 1, 2, and 3 are externally tangent to each other and internally tangent to a circle of radius 6, as shown. The quadrilateral connecting the centers of these four circles is a rectangle. What is the perimeter of this rectangle? | 6-2.

6-3. In how many different ways can 3 men and 3 women line up in a row so that no two people of the same gender stand adjacent to each other? | 6-3.

6-4. What are both values of x for which $\sqrt{\log x}$, $\sqrt{\log 3}$, and $\sqrt{\log 4}$ could be the lengths of the sides of a right triangle? | 6-4.

6-5. Line ℓ is perpendicular to lines m and n. If the product of the slopes of all 3 lines is -8, what is the slope of line ℓ? | 6-5.

6-6. In this problem, brackets will denote the greatest integer function, so that $[n]$ is the greatest integer $\leq n$. For example, $[\pi] = 3$. What value of x satisfies $x^{[x]} = \sqrt[4]{2014}$? [NOTE: For this problem, your answer must be exact. A decimal approximation is **NOT** acceptable.] | 6-6.

HIGH SCHOOL MATHEMATICS CONTESTS

Math League Press, P.O. Box 17, Tenafly, New Jersey 07670-0017

Contest Number 1 *Any calculator without a QWERTY keyboard is allowed.* Answers must be exact *or* have 4 (or more) significant digits, correctly rounded. **October 14, 2014**

Name _____ Teacher _____ Grade Level ____ Score ____

Time Limit: 30 minutes   *Answer Column*

1-1. The lengths of the sides of a rectangle are the positive numbers x, x^2, x^3, and x^4. What is the area of this rectangle?	1-1.
1-2. What is the larger of the only two primes n for which $\dfrac{2\,326\,045}{n}$ has a prime value?	1-2.
1-3. What is the only value of n that satisfies $$2014^n + 2013 \times 2014^{2012} + 2013 \times 2014^{2013} = 2014^{2014}?$$	1-3.
1-4. What is the sum of the degree-measures of the angles at the outer points A, B, C, D, and E of a five-pointed star, as shown?	1-4.
1-5. What is the ordered pair of positive integers (k,b), with the least value of k, which satisfies $\sqrt{2} \cdot \sqrt[3]{3} \cdot \sqrt[4]{4} = \sqrt[k]{b}$?	1-5.
1-6. A face-down stack of 8 playing cards consisted of 4 Aces (A's) and 4 Kings (K's). After I revealed and then removed the top card, I moved the new top card to the bottom of the stack without revealing the card. I repeated this procedure until the stack was left with only 1 card, which I then revealed. The cards revealed were $AKAKAKAK$, in that order. If my original stack of 8 cards had simply been revealed one card at a time, from top to bottom (without ever moving cards to the bottom of the stack), in what order would they have been revealed?	1-6.

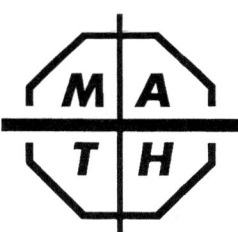

HIGH SCHOOL MATHEMATICS CONTESTS

Math League Press, P.O. Box 17, Tenafly, New Jersey 07670-0017

Contest Number 2   *Any calculator without a QWERTY keyboard is allowed.* Answers must be exact *or* have 4 (or more) significant digits, correctly rounded.   **November 11, 2014**

Name _____ Teacher _____ Grade Level _____ Score _____

Time Limit: 30 minutes   *Answer Column*

2-1. One circle is internally tangent to, and passes through the center of, a larger circle. If the area of the larger circle is 16π, what is the area of the smaller circle?

2-1.

2-2. For positive integral values of n, the symbol $n!$ means the product of the first n positive integers. What positive integer n satisfies

$$n! = \frac{2015! - 2014!}{2014}?$$

2-2.

2-3. If $x > 0$ and I divide x by y, the quotient is 3 and the remainder is 7. If I divide y by x, the remainder is 12. What is the value of x?

2-3.

2-4. For what value of a is one root of $x^2-(2a+1)x+a^2+2 = 0$ twice the other root?

2-4.

2-5. Each time I withdrew \$32 from my magical bank account, the account's remaining balance doubled. *No other account activity was permitted.* My fifth \$32 withdrawal caused my account's balance to become \$0. With how many dollars did I open that account?

2-5.

2-6. In how many ways can I select six of the first 20 positive integers, disregarding the order in which these six integers are selected, so that no two of the selected integers are consecutive integers?

2-6.

HIGH SCHOOL MATHEMATICS CONTESTS

Math League Press, P.O. Box 17, Tenafly, New Jersey 07670-0017

Contest Number 3   *Any calculator without a QWERTY keyboard is allowed.* Answers must be exact *or* have 4 (or more) significant digits, correctly rounded.   **December 9, 2014**

Name _____ Teacher _____ Grade Level ____ Score ____

Time Limit: 30 minutes   *Answer Column*

3-1. Rectangle R is partitioned into 4 small congruent rectangles, as shown. If the perimeter of one of these small rectangles is 2014, what is the perimeter of R?

3-1.

3-2. If $x^3 + y^3 = 200$ and $xy(x+y) = 100$, what is the value of $(x+y)^3$?

3-2.

3-3. Daffie Duck flew from A to B at 30 km per hour. Happie Hawk flew the same route at 45 km per hour. Happie started 2 hours after Daffie, but arrived 1 hour sooner. What is the flight-distance from A to B, in km?

3-3.

3-4. If, for all real x, $f(x) = 2^x f(1-x)$, what is the numerical value of $f(3)$?

3-4.

3-5. What are all real numbers x for which an acute triangle can have side-lengths x, $x+1$, and 5?

3-5.

3-6. How many pairs of positive integers (without regard to order) have a least common multiple of 540?

3-6.

HIGH SCHOOL MATHEMATICS CONTESTS

Math League Press, P.O. Box 17, Tenafly, New Jersey 07670-0017

Contest Number 4 *Any calculator without a QWERTY keyboard is allowed.* Answers must be exact *or* have 4 (or more) significant digits, correctly rounded. **January 13, 2015**

Name _____ Teacher _____ Grade Level _____ Score _____

Time Limit: 30 minutes   NEXT CONTEST: FEB. 10, 2015   *Answer Column*

4-1. For what value of x is 2015 midway between $x-2015$ and $x+2015$?

4-1.

4-2. Right triangles I and II are *not* congruent, but two sides of $\triangle$I are congruent to two sides of $\triangle$II. If the side-lengths of $\triangle$I are 3, 4, and 5, what is the least possible length of the hypotenuse of $\triangle$II?

4-2.

4-3. What is the smallest positive integer x for which $5x^{301} + 4x^{300}$ is the cube of an integer?

4-3.

4-4. If the square of the smaller of two consecutive positive integers is x, what is the square of the larger of these two integers, in terms of x?

4-4.

4-5. A pair of salt and pepper shakers comes in two types: identical and fraternal. Identical pairs are always the same color. Fraternal pairs are the same color half the time. The probability that a pair of shakers is fraternal is p and that a pair is identical is $q = 1-p$. If a pair of shakers is of the same color, determine, in terms of the variable q alone, the probability that the pair is identical.

4-5.

4-6. As shown, one angle of a triangle is divided into four smaller congruent angles. If the lengths of the sides of this triangle are 84, 98, and 112, as shown, how long is the segment marked x?

84   112

98   x

4-6.

Contest Number 5   *Any calculator without a QWERTY keyboard is allowed.* Answers must be exact *or* have 4 (or more) significant digits, correctly rounded.   **February 10, 2015**

Name _____ Teacher _____ Grade Level _____ Score _____

Time Limit: 30 minutes   NEXT CONTEST: MAR. 17, 2015   *Answer Column*

5-1. What value of x satisfies $(x-2014)(x+2015) = (x+2014)(x-2015)$?

5-1.

5-2. The slope and y-intercept of line ℓ are equal to each other and are reciprocals of each other. What is the x-intercept of line ℓ?

5-2.

5-3. What is the perimeter of an equilateral triangle in which the distance from the center to each vertex is 1, as shown?

5-3.

5-4. If a solution of the equation $4^x = 2^x+6$ can be expressed as $\log_b n$, where b and n are both primes, what is the ordered pair (b,n)?

5-4.

5-5. How long is the longer diagonal of a rhombus whose perimeter is 60, if three of its vertices lie on a circle whose diameter is 25, as shown?

5-5.

5-6. The 14 cabins of the Titanic Mail Boat are numbered consecutively from 1 through 14, as are the 14 room keys. In how many different ways can the 14 room keys be placed in the 14 rooms, 1 per room, so that, for every room, the sum of that room's number and the number of the key placed in that room is a multiple of 3?

5-6.

HIGH SCHOOL MATHEMATICS CONTESTS

Math League Press, P.O. Box 17, Tenafly, New Jersey 07670-0017

Contest Number 6 *Any calculator without a QWERTY keyboard is allowed.* Answers must be exact *or* have 4 (or more) significant digits, correctly rounded. **March 17, 2015**

Name _____ Teacher _____ Grade Level ____ Score ____

Time Limit: 30 minutes | *Answer Column*

6-1. What is the greatest possible length of the longest side of an isosceles triangle whose sides each have an integral length and whose perimeter is 2015?

6-1.

6-2. Pat's coins consist of only pennies, nickels, dimes, and quarters. Pat has fewer than 100 pennies, and the total value of all of Pat's coins is $7.21. At most how many pennies does Pat have?

6-2.

6-3. As shown, the centers of two circles are endpoints of a line segment of length 50. If the radii of the two circles have lengths 30 and 40, how long is the part of the line segment that lies in both circles?

6-3.

6-4. If x and y are real numbers whose sum is three times their product, and whose product is not 0, what is the value of $\frac{1}{x} + \frac{1}{y}$?

6-4.

6-5. For some constant b, if the minimum value of $f(x) = \frac{x^2 - 2x + b}{x^2 + 2x + b}$ is $\frac{1}{2}$, what is the maximum value of $f(x)$?

6-5.

6-6. If the lengths of two sides of a triangle are $60\cos A$ and $25\sin A$, what is the greatest possible integer-length of the third side?

6-6.

Solutions on Page 57 • Answers on Page 67

HIGH SCHOOL MATHEMATICS CONTESTS

Math League Press, P.O. Box 17, Tenafly, New Jersey 07670-0017

Contest Number 1 *Any calculator without a QWERTY keyboard is allowed.* Answers must be exact *or* have 4 (or more) significant digits, correctly rounded. **October 13, 2015**

Name _____ Teacher _____ Grade Level ____ Score ____

Time Limit: 30 minutes   *Answer Column*

1-1. Each vertex of a square is assigned a different positive integer. If the numbers on the endpoints of each diagonal have the same sum, what is the least possible value of this sum?

1-1.

1-2. Every coin in my piggy bank has a face value of 50¢, 25¢, 10¢, 5¢, or 1¢. The bank contains many coins of each type. At most how much money can I withdraw from my piggy bank without being able to make change for $1?

1-2.

1-3. What are all integers x for which $|x^2 - 26x + 88|$ is a prime?

1-3.

1-4. In the big right triangle shown, the lengths of the legs are 8 and 15. How long is the line segment whose length is marked x?

8, 6, x, 9

1-4.

1-5. Which integer > 1 leaves the same remainder when divided into each of the numbers 1108, 1453, 1844, and 2281?

1-5.

1-6. What is the largest value of c for which exactly three different pairs of positive integers (x,y) satisfy $5x + 7y = c$?

1-6.

Solutions on Page 58 • Answers on Page 67

HIGH SCHOOL MATHEMATICS CONTESTS

Math League Press, P.O. Box 17, Tenafly, New Jersey 07670-0017

Contest Number 2 *Any calculator without a QWERTY keyboard is allowed.* Answers must be exact *or* have 4 (or more) significant digits, correctly rounded. **November 10, 2015**

Name _____ Teacher _____ Grade Level _____ Score ____

Time Limit: 30 minutes | *Answer Column*

2-1. If 23 is written as the sum of the squares of 4 positive integers (not necessarily different), what is the largest square in this sum?	2-1.
2-2. One side of an equilateral triangle is the diameter of a circle, and one vertex of the triangle lies on a larger circle, concentric with the smaller circle, as shown. If the area of the smaller circle is 16π, what is the area of the larger circle?	2-2.
2-3. When ten mathletes huddled together, they spaced themselves equally around a circle. The numbers on their uniforms were not necessarily distinct, and their sum was 300. If each number was the average of the two numbers nearest it, what was the largest of the ten numbers on their uniforms?	2-3.
2-4. What are all real values of $x \neq 0$ that satisfy $|x|^{x^2-x-2} < 1$?	2-4.
2-5. What is the smallest integer $x > 1$ for which $\sqrt{x\sqrt{x\sqrt{x}}}$ is an integer?	2-5.
2-6. The length of each side of a triangle is the reciprocal of a different integer. If one of these integers is 2015, what is the least possible sum of the other two integers?	2-6.

© 2015 by Mathematics Leagues Inc.

Solutions on Page 59 • Answers on Page 67

HIGH SCHOOL MATHEMATICS CONTESTS

Math League Press, P.O. Box 17, Tenafly, New Jersey 07670-0017

Contest Number 3 *Any calculator without a QWERTY keyboard is allowed.* Answers must be exact *or* have 4 (or more) significant digits, correctly rounded. **December 8, 2015**

Name _____ Teacher _____ Grade Level ____ Score ____

Time Limit: 30 minutes | *Answer Column*

3-1. What is the largest possible degree-measure of an angle of a triangle if the degree-measures of all three angles are positive integers? | 3-1.

3-2. A magical hat takes any number fed into it and divides 1492 by that number. If 2015 is fed into the machine and the first output is fed back into the machine, what is the value of the second output? | 3-2.

3-3. In a certain sequence of numbers, each number after the first is the sum of all the preceding numbers. If this sequence's 100th term is 2015, what is its 101st term? | 3-3.

3-4. Twelve congruent isosceles triangles share a common vertex, as shown. If the sum of the measures of all the angles that share the common vertex is 360°, what is the measure of each triangle's smallest angle? | 3-4.

3-5. For what integer $k > 0$ can $(\sqrt{2}-1)^5$ be written as $\sqrt{k+1} - \sqrt{k}$, the difference between the square roots of two consecutive integers? | 3-5.

3-6. There are an infinite number of ordered pairs of positive integers (m,n) such that $m^3 = n^2$ and $m + n$ is a perfect square. One such pair is $(m,n) = (9,27)$. What is the largest value of $m < 1000$ for which such an ordered pair exists? | 3-6.

HIGH SCHOOL MATHEMATICS CONTESTS

Math League Press, P.O. Box 17, Tenafly, New Jersey 07670-0017

Contest Number 4   *Any calculator without a QWERTY keyboard is allowed.* Answers must be exact *or* have 4 (or more) significant digits, correctly rounded.   **January 12, 2016**

Name _____ Teacher _____ Grade Level _____ Score _____

Time Limit: 30 minutes | *Answer Column*

4-1. For all real numbers x, the function f is defined by $f(x) = 2016$. What is the value of $f(x+2016)$? | 4-1.

4-2. What is the sum of the areas of the five shaded triangles shown at the right that are drawn interior to a 3 by 6 rectangle? | 4-2.

4-3. If $A^{2x} = 4$ and $A > 0$, what is the numerical value of $\dfrac{A^{3x} - A^{-3x}}{A^{x} - A^{-x}}$, written as a ratio of positive integers in lowest terms? | 4-3.

4-4. Al runs three times as fast as he walks. It takes Al 21 minutes to get to work from home if he walks for twice the amount of time that he runs. How many minutes does it take Al to get to work from home if he runs for twice the amount of time that he walks? | 4-4.

4-5. At most how many of the first 100 positive integers can be chosen if no two of the chosen numbers have a sum divisible by 5? | 4-5.

4-6. What is the area of quadrilateral $ABCD$ whose vertices have polar coordinates $A(0,0)$, $B(4,0)$, $C(3,\frac{\pi}{8})$, $D(1,\frac{3\pi}{8})$? | 4-6.

HIGH SCHOOL MATHEMATICS CONTESTS

Math League Press, P.O. Box 17, Tenafly, New Jersey 07670-0017

Contest Number 5 *Any calculator without a QWERTY keyboard is allowed.* Answers must be exact *or* have 4 (or more) significant digits, correctly rounded. **February 9, 2016**

Name _____ Teacher _____ Grade Level ____ Score ____

Time Limit: 30 minutes | *Answer Column*

5-1. If $-1 < x < 0$, then for what positive integer $n \le 2016$ does x^n take on its least value?

5-1.

5-2. If we remove the first + sign and then "close" the resulting space in $1+2+3+4+5+6+7+8+9$, we'll get $12+3+4+5+6+7+8+9 = 54$. What is the ordinal number (first, second, etc.) of the + sign whose removal from $1+2+3+4+5+6+7+8+9$, followed by a "closing" of the resulting space, would make the resulting sum equal 99?

5-2.

5-3. For how many different integers $b > 1$ is $\log_b 256$ a positive integer?

5-3.

5-4. In the diagram, three segments are drawn interior to a square. Each segment connects a vertex of the square to the midpoint of a side of the square. If the area of the larger shaded triangle is 150, what is the area of the smaller shaded triangle?

5-4.

5-5. Regular polygons M and N have m and n sides respectively, with $m > n$. What are all ordered pairs (m,n) for which the ratio of the measure of an interior angle of M to the measure of an interior angle of N is 3:2?

5-5.

5-6. From a box containing 20 gold, 10 silver, and some bronze medals, the probability of randomly selecting 2 gold and 2 silver medals, with replacement, equals the probability of randomly selecting 1 gold, 1 silver, and 2 bronze medals, with replacement. How many bronze medals are in the box?

GOOD WORK!

5-6.

HIGH SCHOOL MATHEMATICS CONTESTS

Math League Press, P.O. Box 17, Tenafly, New Jersey 07670-0017

Contest Number 6 *Any calculator without a QWERTY keyboard is allowed.* Answers must be exact *or* have 4 (or more) significant digits, correctly rounded. **March 15, 2016**

Name _____ Teacher _____ Grade Level _____ Score _____

Time Limit: 30 minutes | *Answer Column*

6-1. What is the least possible area of each of two rectangles which have equal areas and integral side-lengths, but are not congruent?

6-1.

6-2. What is the length of the hypotenuse of a right triangle, whose legs have lengths $\frac{\pi}{3}$ and $\frac{\pi}{4}$?

6-2.

6-3. For what value of k do $x^3+kx^2-3x+4 = 0$ and $x^3+kx^2-5x+8 = 0$ have a common solution?

6-3.

6-4. In polygon P at the right, every angle is 45°, 90°, or 135°. If the length of every segment is 1, what is the area of P?

6-4.

6-5. If $0° < x \le 2016°$, how many angles x satisfy
$$\sin^2 2016° + \sin^2 x = 1?$$

6-5.

6-6. Pat and Lee alternately toss two fair dice, Pat tossing first. The first to roll a 9 wins the money paid by both players. If Lee pays $400 to play, how many dollars should Pat pay to make this game fair? [In a *fair game*, each player's expected value is 0 (no net gain or loss).]

6-6.

Solutions on Page 63 • Answers on Page 67

31

Complete Solutions
October, 2011 – March, 2016

Problem 1-1

By observation, the solution is $\sqrt{1} + \sqrt{1} = 1 + 1 = 2 = \sqrt{4}$, so $(m,n) = \boxed{(1,1)}$.

Problem 1-2

Method I: Subtract $(x-2011)(x-2012)$ from both sides and factor to get $(x-2011)(x-2010-x+2012) = 0 \Leftrightarrow 2(x-2011) = 0$. The $\boxed{1}$ solution is $x = 2011$.

Method II: If $x = 2011$, both sides are 0. If $x \neq 2011$, divide both sides by $(x-2011)$ to get $x-2010 = x-2012$, which has no solution. Since the only solution is $x = 2011$, the number of solutions is 1.

Problem 1-3

The first time around the road, the letters to which the rabbit hops are $A, D, G, J, \ldots, Y$. The rabbit continues, landing on $B, E, H, K, \ldots, Z$. Finally, the third time around the road, the rabbit lands on the remaining letters in the order $C, F, I, L, \ldots, X$. The last letter to which the rabbit hopped was $\boxed{X}$.

Problem 1-4

Method I: An altitude to the hypotenuse of a right triangle splits the right triangle into two right triangles that are similar to each other and to the original right triangle, the one with perimeter $\boxed{60}$, as shown.

Method II: The altitude to the hypotenuse is the mean proportional between the segments of the hypotenuse. Look for two integers whose product is 12^2. Continue as in the diagram above.

[**NOTE:** The only 4 right triangles with a leg of length 12 and integer side-lengths have side-lengths 5-12-13, 9-12-15, 12-16-20, and 12-35-37.]

Problem 1-5

Make a table of earnings, where the number of dollars earned on days 1 and 2 are x and y, respectively, and on each succeeding day, the number of dollars earned is the sum of the earnings of the previous 2 days:

Day	1	2	3	4	5	6
\$	x	y	$x+y$	$x+2y$	$2x+3y$	$3x+5y$

Day	7	8	9	10
\$	$5x+8y$	$8x+13y$	$13x+21y$	$21x+34y$

The only positive integral solution of $21x+34y = 55$ is $(1,1)$, so day 3 earnings were $x+y = 1+1 = \boxed{2}$.

Problem 1-6

Method I: The sum is divisible by 7 if and only if $2^x + 3^x = 7y$ for some integer y. To use a calculator, first solve this equation for y. Then, you can use the calculator's table function to evaluate $y = (2^x + 3^x)/7$ starting at $x = 1$. Look for a pattern. For which integers x is y an integer? The first few such values of x are $x = 3, 9, 15, 21, 27, \ldots$. These are all odd multiples of 3. Thus, the x we want is the smallest number bigger than 1 million that's an odd multiple of 3. It's $\boxed{1\,000\,005} = 10^6 + 5$.

Method II: Use the identities $2^{3n} = 8^n = (7+1)^n$, and $3^{3n} = 27^n = (4 \times 7 - 1)^n$. Expand and add terms to get that $2^{3n} + 3^{3n}$ is divisible by 7 whenever n is an odd number. Here's why: Except for the constant term in either binomial expansion, every term is a multiple of 7. When n is odd, the constants are 1 and -1. Their sum is 0. When n is even, the constants are both 1, leaving a remainder of 2 when the sum is divided by 7. (Similarly, if the exponents were $3n+1$ or $3n+2$, the constant terms would **not** drop out.) The sum is divisible by 7 if and only if the exponent is an odd multiple of 3. Clearly, the smallest odd multiple of 3 that is greater than 1 million is $1\,000\,005 = 10^6 + 5$.

Contests written and compiled by Steven R. Conrad & Daniel Flegler   Mathematics Leagues Inc., © 2011

Problem 2-1

The expressions within the parentheses are opposites, so the sum of their cubes is always $\boxed{0}$.

Problem 2-2

Every line through the center of the square will divide the square into two regions of equal area. The line shown in the diagram is the required line. This line passes through (0,0) and (4,1), and it has a slope of $\boxed{\frac{1}{4}}$ or 0.25.

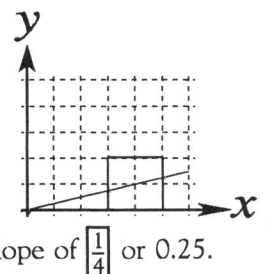

Problem 2-3

Let the integers be x and $x+1$. Their squares will always differ by $2x+1$, which is also the sum of the two integers. Since the difference of the squares is 2011, the sum of the integers must also be $\boxed{2011}$.

Problem 2-4

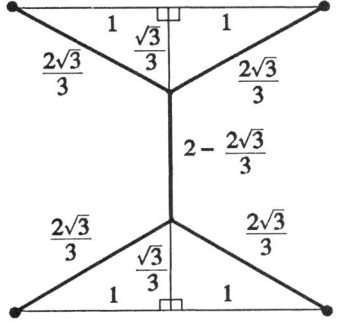

The total Y-length is

$$4\left(\frac{2\sqrt{3}}{3}\right) + 2 - \frac{2\sqrt{3}}{3} =$$

$\boxed{2+2\sqrt{3}}$.

[**NOTE:** The sum of the lengths of the square's diagonals is $4\sqrt{2} = 5.6568542\ldots$, while the length of the double Y is $2 + 2\sqrt{3} = 5.4641016\ldots$.]

Problem 2-5

The positive integer N has 5 digits, so the value of $N1$ is $10N + 1$, and the value of $1N$ is $100\,000 + N$. From the information given, we can write the equation $10N + 1 = 3(100\,000 + N)$. Solve this equation to get $N = \boxed{42\,857}$.

Problem 2-6

Expressing 1 as a sum of the indicated type is similar to expressing 1/8 as a sum of decimals that consist of only 0s and 1s. We will multiply these decimals by 8 to answer problem 2-6. Since $1/8 = 0.125$, we know that $1 = 8 \times 0.125$. If we split $1/8 = 0.125$ into decimals that use only the digits 0 and 1, we can multiply by 8 to split 1 into decimals that use only the digits 0 and 8. Let's get the 1s as far left as possible so we will use as few such decimals as possible. Begin by writing $0.125 = 0.111 + 0.011 + 0.001 + 0.001 + 0.001$. Multiplying by 8, $1 = 0.888 + 0.088 + 0.008 + 0.008 + 0.008$. We need at least $\boxed{5}$ such decimals.

Contests written and compiled by Steven R. Conrad & Daniel Flegler

Problem 3-1

If two positive numbers have a fixed product, their sum will be minimized when the numbers are as close together as possible. Thus, the smallest perimeter of such a rectangle is $4 \times 5 = 20$, the perimeter of a square. The largest such perimeter is that of a 1×25 rectangle. Its perimeter is 52. The difference between the perimeters is $52 - 20 = \boxed{32}$.

Problem 3-2

Since 8 is the only mode and 5 is the median, there must be two 8s and one 5 among the 5 integers. The mean of the 5 integers is 5, so their sum is $5 \times 5 = 25$. Since $8 + 8 + 5 = 21$, the two missing numbers must total 4 and average 2. Since 8 is the only mode, there cannot be two 2s. Since all numbers are positive integers, the five numbers must be $\boxed{1, 3, 5, 8, 8}$.

Problem 3-3

Intuitively, this product will be maximized whenever $\tan x = \tan y$, *i.e.*, when $x = y = 30°$. At this point, $(\tan x)(\tan y) = \left(\frac{1}{\sqrt{3}}\right)\left(\frac{1}{\sqrt{3}}\right) = \boxed{\frac{1}{3}}$.

PROOF: Let's verify that the maximum possible product actually is $\frac{1}{3}$. Let $t = \tan x$; so $0 < t < \sqrt{3}$. Is $\tan x \tan y = \tan x \tan(60° - x) = t\left(\frac{\sqrt{3} - t}{1 + t\sqrt{3}}\right) \le \frac{1}{3}$ a valid inequality? Since t is positive, we can clear fractions without concern. Clearing fractions, we get $3t\sqrt{3} - 3t^2 \le 1 + t\sqrt{3}$. That's true precisely when $t^2 - \frac{2t}{\sqrt{3}} + \frac{1}{3}$ is non-negative. This last expression is a perfect square—it's $\left(t - \frac{1}{\sqrt{3}}\right)^2$—so the expression is always non-negative. Reversing these steps proves that the maximum value guessed in Method I is correct.

Problem 3-4

Flash and Ace tie if Ace runs 75 m while Flash runs 60 m, so Flash runs at $\frac{4}{5}$ of Ace's speed. Continuing, Speedy runs at $\frac{4}{5}$ of Flash's speed, or $\frac{4}{5} \times \frac{4}{5} = \frac{16}{25} = \frac{48}{75}$ of Ace's speed. In meters, the head start needed by Speedy is $75 - 48 = \boxed{27}$.

Problem 3-5

If $f(x) = ax + b$ and $g(x) = cx + d$, then $g(f(x)) = c(ax + b) + d = acx + (d + bc) = 2x + 6$. Thus, $ac = 2$. Also, $f(g(x)) = a(cx + d) + b = acx + (b + ad)$. Since its graph passes through the origin, the constant term, $b + ad$, must equal 0. Therefore, $f(g(x)) = 2x$ and $f(g(2011)) = 2(2011) = \boxed{4022}$.

[**NOTE:** There are only 3 such function pairs $f(x)$ and $g(x)$ with integral coefficients. We must satisfy $ac = 2$, $d + bc = 6$, and $b + ad = 0$. Solving, $b = -ad$, so $d + (-ad)c = d(1 - ac) = d(1 - 2) = 6$, so $d = -6$. The only 3 pairs of functions $(f(x), g(x))$ that satisfy the requirements in the second sentence of this note are $(2x + 12, x - 6)$, $(-x - 6, -2x - 6)$, $(-2x - 12,$ and $-x - 6)$.]

Problem 3-6

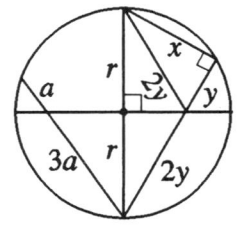

By the Pythagorean Theorem, $x^2 + (3y)^2 = (2r)^2$. Expanding, $x^2 + 9y^2 = 4r^2$. Furthermore, $x^2 + y^2 = (2y)^2$, or $x^2 = 3y^2$. By substitution, $12y^2 = 4r^2$, so $y = \frac{r}{\sqrt{3}}$. Thus, $3y = r\sqrt{3} = $ the length of the longer chord. Substituting into $x^2 = 3y^2$, we see that the length of the shorter chord is $\frac{4r}{\sqrt{6}}$. Hence, the ratio of their chord-lengths, longer to shorter, is $\boxed{\frac{3\sqrt{2}}{4}}$.

Contests written and compiled by Steven R. Conrad & Daniel Flegler ©2011 by Mathematics Leagues Inc.

Problem 4-1

The average of the integers is their sum divided by the number of integers $= \frac{2012^3}{2012} = \boxed{2012^2}$.

Problem 4-2

Method I: Since $\frac{3}{90} < \frac{3}{89} < \frac{3}{87}$, we have $\frac{1}{30} < \frac{3}{89} < \frac{1}{29}$. Therefore, the fractions are $\boxed{\frac{1}{29}, \frac{1}{30}}$.

Method II: Using a calculator, we get $\frac{3}{89} \approx 0.0337$. Since $\frac{1}{0.0337} \approx 29.67$, the fractions are $\frac{1}{29}$ and $\frac{1}{30}$.

Method III: $\frac{3}{89} \approx \frac{1}{29.67}$, so $\frac{1}{30} < \frac{3}{89} < \frac{1}{29}$.

Problem 4-3

If none of the three consecutive months were shorter than 30 days long, then these months would contain at least $30+30+31 = 91 = 7 \times 13$ days, which is exactly 13 weeks. To have only 12 Fridays, we'd need to include a month with fewer than 30 days. Taken together, the months December, January, and February; or January, February, and March; or February, March, and April have 88, 89, or 90 days. My birthday must fall in the same month no matter which set of 3 months has exactly 12 Fridays. Therefore, my birthday must fall in $\boxed{\text{February}}$.

Problem 4-4

Method I: Draw $\overline{AC}$. Since $\overline{AB}$ and $\overline{AC}$ are both diagonals of 1×2 rectangles, $AB = AC$. The slope of $\overline{AB}$ is $\frac{1}{2}$, and the slope of $\overline{AC}$ is -2, so $\overline{AB} \perp \overline{AC}$. Hence, $\triangle BAC$ is an isosceles right triangle, and $m\angle ABC = \boxed{45}$ or 45°.

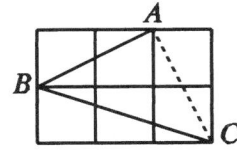

Method II:
$m\angle ABC = m\angle ABE + m\angle DBC$
$= \text{Arctan } \frac{1}{2} + \text{Arctan } \frac{1}{3} = \text{Arctan } 1 = 45$.

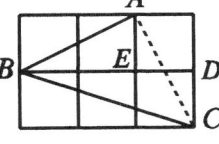

Method III: The lengths of $\triangle ABC$'s 3 sides are $\sqrt{5}$, $\sqrt{5}$, and $\sqrt{10}$, so it is an isosceles right triangle.

Problem 4-5

Let P be a polynomial of degree n. Then $P(x+5) - P(x)$ cannot be constant unless $n = 1$ or 0. But $n \neq 0$ (since that would make $P(x+5) - P(x) = 0$), so $n = 1$ and P is linear. Let $P(x) = mx + b$. Then $2 = P(x+5) - P(x) = 5m$. Therefore, $m = \frac{2}{5}$ and $P(x) = \frac{2}{5}x + b$. Now calculate $P(4) - P(2) = \frac{2}{5}(2) = \boxed{\frac{4}{5}}$.

Problem 4-6

Method I: $x^2 - 4xy + 5y^2 + 2x - 12y + 17 = 0 \Leftrightarrow$
$x^2 - 4xy + 4y^2 + y^2 + 2x - 12y + 17 = 0 \Leftrightarrow$
$(x-2y)^2 + 2(x-2y) + 1 + y^2 - 8y + 16 = 0 \Leftrightarrow$
$(x - 2y + 1)^2 + (y - 4)^2 = 0 \Rightarrow$
$(x,y) = \boxed{(7,4)}$. This solution checks.

Method II: If we treat the first equation as a quadratic in x, that equation will become $x^2 + x(-4y + 2) + (5y^2 - 12y + 17) = 0$. Since the discriminant of this equation is $-4(y-4)^2$, any real solution requires that $y = 4$, and so on.

Method III: Subtract the second equation from the first to get $4xy - 4y^2 - 16x + 4y + 48 = 0$. This has no x^2 term. Let's solve it for x. $x(y-4) = y^2 - y - 12 = (y-4)(y+3)$. If $y = 4$, substitution into either of the original equations yields $x = 7$. If $y \neq 4$, then $x = y+3$. Substitute into the first equation to get $2(y-4)^2 = 0$, a contradiction. The one real solution is $(7,4)$.

[**NOTE:** If you eliminate xy by multiplying the first equation by 2 and subtracting the second equation, you will get $x^2 + y^2 - 14x - 8y + 65 = (x-7)^2 + (y-4)^2 = 0$. The problem is this method *doesn't* always yield a solution. For example, if $2x^2 - 7xy + 3y^2 + 11x - 5y - 7 = 0$ were the 2nd equation, then (7,4) would be a solution, but not by this method.]

Contests written and compiled by Steven R. Conrad & Daniel Flegler   ©2012 by Mathematics Leagues Inc.

Contest # 5   *Answers & Solutions*   **2/14/12**

Problem 5-1

We're told that $\sqrt{\pi r^2} = 2012\pi$, so $\pi r^2 = 2012^2\pi^2$. Solving, $r = \boxed{2012\sqrt{\pi}}$.

Problem 5-2

Since $\sqrt{600} \approx 24.5$, the consecutive integers must be -24 and -25. Or, $x^2+kx+600 = (x+24)(x+25) = (x-24)(x-25)$. For the roots to be consecutive negative integers, choose the first factoring. The value of k is $\boxed{49}$.

Problem 5-3

We are told that $x^n = x^{n-1} + x^{n-2}$. Since $x \neq 0$, we can divide by x^{n-2} to get $x^2 = x+1$. Thus, the only possible values of x are $\boxed{\frac{1 \pm \sqrt{5}}{2}}$.

Problem 5-4

If x is the dollar cost of one of the lovebirds, then the loss is $x-24$. This loss is also $x\%$ of the cost, so this loss is also $\frac{x}{100}(x) = \frac{x^2}{100}$. When we set these two different expressions for the loss equal to each other to get $\frac{x^2}{100} = x-24$, or $x^2-100x+2400 = (x-60) \times (x-40) = 0$, the two possible values of x are $\boxed{40, 60}$.

Problem 5-5

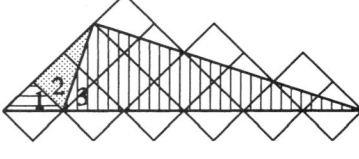

By right triangle trigonometry, $\tan\angle 1 = \frac{1}{1}$ and $\tan\angle 2 = \frac{2}{1}$. Hence, $m\angle 1 = \text{Arctan } 1$ and $m\angle 2 = \text{Arctan } 2$. What is $m\angle 3$? The shorter leg of the largest of the 3 shaded right triangles is the diagonal of a 1×2 rectangle. The longer leg is exactly 3 times as long as the shorter leg since it passes through 3 of the 1×2 rectangles. Therefore, $m\angle 3 = \text{Arctan } 3$. Hence $(a,b,c) = \boxed{(1,2,3)}$.

Problem 5-6

Let the roots be r, r, and s. Since the coefficient of x^2 is 0, the sum of the roots $= 2r+s = 0$, so $s = -2r$. The value of the constant term is 1, so the product of the roots is $-1 = r^2 s = -2r^3$. Solving, $r = \boxed{\sqrt[3]{\frac{1}{2}}}$.

Contests written and compiled by Steven R. Conrad & Daniel Flegler   © 2012 by Mathematics Leagues Inc.

Problem 6-1

The other point must be the reflection, through the origin, of the point (−2011,−2012). The co-ordinates of that reflection point are (2011,2012).

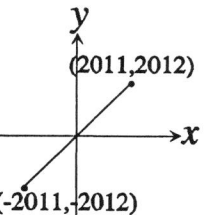

Problem 6-2

Since the sum of 3 odd primes would be odd, one of the primes is 2. The only two primes whose sum is 38 are 7 and 31. Since Dad is the oldest of the three, his age was 31.

Problem 6-3

$(\sqrt{a}+\sqrt{b}+\sqrt{c})^2 = a+b+c+2\sqrt{ab}+2\sqrt{ac}+2\sqrt{bc} = 10+\sqrt{24}+\sqrt{40}+\sqrt{60} = 10+2\sqrt{6}+2\sqrt{10}+2\sqrt{15}$. Now, $ab = 6 = 2\times3$, $ac = 10 = 2\times5$, and $bc = 15 = 3\times5$, so $(a,b,c) = (2,3,5)$.

Problem 6-4

If a number equals its reciprocal, either both equal 1 or both equal −1. Therefore, if the function passes

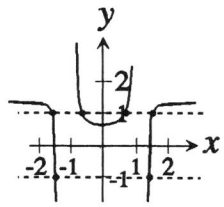

through a point with a y-coordinate of 1 or −1, its reciprocal will also pass through the very same point. Hence, all we must do is draw $y = 1$ and $y = -1$ and then count the number of points through which these two lines cross the graph of the function. That number, 6, is the number of intersection points of the graphs of the function and its reciprocal.

Problem 6-5

Let's use the laws of logarithms to rewrite the given equation as $[\log_2(x+1)^8][\log_2(x+1)] = 8$. Next, we rewrite this equation as $[8\log_2(x+1)][\log_2(x+1)] = 8 \Leftrightarrow [\log_2(x+1)]^2 = 1$. Therefore, $\log_2(x+1) = \pm1$, from which we get $x+1 = 2$ or $x+1 = 2^{-1} = \frac{1}{2}$. Solving, we get $x = 1$ or $x = -\frac{1}{2}$.

Problem 6-6

Method I: Let G be the number of green balls. If there is a solution in the case that the number of non-green balls is 1, this solution must be minimal, so let's begin by assuming there is a solution in that case. Then $\frac{G}{G+1} \times \frac{G-1}{G} \times \frac{G-2}{G-1} \times \frac{G-3}{G-2} \times \frac{G-4}{G-3} = \frac{G-4}{G+1} = \frac{1}{2}$, so $2G-8 = G+1$, and $G = 9$. Thus, the total number of balls is $9+1 = 10$.

Method II: The smallest number will be achieved when there are as few non-green balls in the bag as possible. Let's see what happens if we let there be only 1 non-green ball. The simplest solution is to note that 5 is half of 10, and that if there are 10 balls altogether, and only 1 of them is not green, then we can split them into two piles, each containing 5 balls. One pile is all-green, and one has a single non-green ball. In that case, we'll pick either pile with probability one-half. Hence, the least number of balls is 10.

Contests written and compiled by Steven R. Conrad & Daniel Flegler   ©2012 by Mathematics Leagues Inc.

Problem 1-1

The ratio of the circumference of the circle to the perimeter of the square is $\frac{2\pi r}{4r} = \boxed{\frac{\pi}{2}}$.

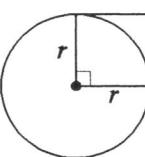

Problem 1-2

The equation $|a-b| = 1005$ means that, on a number line, the distance between points a and b is 1005. Similarly, since $|b-c| = 1007$, the distance between points c and b is 1007. If a and c are on the same side of b, then a and c are 2 units apart. If a and c are on opposite sides of b, then the distance between a and c would be 1005 + 1007 = 2012. Therefore, the two possible values of $|a-c|$ are $\boxed{2, 2012}$.

Problem 1-3

The length of any side of a polygon must be less than half the perimeter. For example, in a triangle with perimeter 100, the sides can have lengths 25, 26, and 49. In a quadrilateral with perimeter 100, the sides can have lengths 17, 17, 17, and 49. No matter how many sides a polygon has, the longest side's length cannot equal or exceed the sum of the lengths of all the remaining sides. The maximum length is $\boxed{49}$.

Problem 1-4

Every positive prime except 2 is odd. The difference between any two odd numbers is even, so two primes can differ by an odd number only if one of the primes is 2. Begin by looking for differences between 2 and odd primes: $3-2 = 1$; $5-2 = 3$; $7-2 = 5$; $11-2 = 9$. So, 7 is the smallest unattainable odd number. Let's see which even numbers are expressble as a difference between odd primes: $5-3 = 2$; $7-3 = 4$; $11-5 = 6$. This proves that the least positive integer we cannot write as the difference between positive primes is $\boxed{7}$.

Problem 1-5

Each non-empty subset of $\{1,2,3,4,5,6,7,8,9\}$ that contains 2 or more digits has one such ordering. The total number of non-empty subsets of a 9-element set is 2^9-1. Since this total includes 9 one-digit subsets (which represent 9 one-digit numbers), we must subtract 9, making the answer $2^9-1-9 = \boxed{502}$.

Problem 1-6

Since x is the number of coins given to each boy, and y is the number of coins given to each girl, if T is the total number of coins, then $T = 3x+4y$. We want the largest T for which there is only one solution in positive integers (x,y). First, let's look at an example of how we can get a second solution when a first solution is known: In this case, by either 1) adding 4 to the value of x while subtracting 3 from the value of y, or 2) subtracting 4 from the value of x while adding 3 to the value of y, we're merely adding and subtracting 12 coins, so the value of T does not change. Thus, since (1,5) is a solution when $T = 23$, the pair (5,2) is a second solution. If $x_0 > 4$ and (x_0,y_0) satisfies $T = 3x+4y$, then (x_0-4,y_0+3) also satisfies $T = 3x+4y$, with $x_0-4 > 0$. Similarly, if $y_0 > 3$ satisfies $T = 3x+4y$, then (x_0+4,y_0-3) also satisfies $T = 3x+4y$, with $y_0-3 > 0$. Therefore, (4,3) is the solution with the largest possible values of x and y for which no second solution pair exists. In this case, $T = \boxed{24}$.

Contests written and compiled by Steven R. Conrad & Daniel Flegler

Problem 2-1

Every such 3-digit number *ddd* can be written as $d \times 111 = d \times 3 \times 37$. Since *d* is a digit, $1 \le d \le 9$, so the largest prime divisor is $\boxed{37}$.

Problem 2-2

The average of 5 numbers is their sum divided by 5. Since the sum of these 5 numbers is $(16+16+16) + (6+6) = 60$, their average is $60/5 = \boxed{12}$.

Problem 2-3

The two small right $\triangle$s are similar to each other (and the large right $\triangle$). A radius of the circle is 12. Thus the longer leg

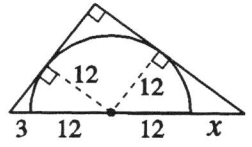

of the right $\triangle$ at the lower left is 12. Since its hypotenuse is 15, its dimensions are 9, 12, 15. The shorter leg of the right $\triangle$ at the lower right is 12, so its dimensions are 12, 16, 20. Since $12+x = 20$, $x = \boxed{8}$.

Problem 2-4

Clearly, $(a,b,c) = (3,3,3)$ is a solution. In any other solution, at least one fraction must exceed $\frac{1}{3}$, which means one fraction must equal $\frac{1}{2}$. Since $0 < a \le b \le c$, it follows that, in any other solution, $a = 2$. Now, solve $\frac{1}{b} + \frac{1}{c} = \frac{1}{2}$ in positive integers. This is a simpler version of the original equation. This time, an obvious solution is $(b,c) = (4,4)$. In any other solution, one fraction must exceed $\frac{1}{4}$. That means that one fraction must equal $\frac{1}{3}$. Thus, $\frac{1}{c} = \frac{1}{2} - \frac{1}{3} = \frac{1}{6}$. Finally, the positive integer solutions are the ordered triples $\boxed{(3,3,3),\ (2,4,4),\ (2,3,6)}$.

Problem 2-5

The integers range from 1 through 100. How many 1's are there? There's $2^{1-1} = 1$ of them. How many 2's? There are $2^1 = 2$ of those. Similarly, there are 2^2 3's, 2^3 4's, ..., 2^{98} 99's. The total number of all these integers is $2^0+2^1+2^2+2^3+ \ldots +2^{98} = 2^{99}-1$. The number of 100's is 2^{99}, so we can pair one 100 with every other integer—and we'll still have one 100 left over. So, if the numbers are ordered from least to greatest, the middle number will be the extra $\boxed{100}$.

Problem 2-6

Place the rectangle on the coordinate axes with vertices (0,0), (2015,0), (2015,2012), and (0,2012). The diagonal is $y = \frac{2012}{2015}x$, with $0 \le x \le 2015$. The key observation is that the diagonal enters a new square each time the diagonal crosses a vertical line of the form $x = a$, with $a = 1, 2, 3, \ldots, 2014$, or a horizontal line of the form $y = b$, with $b = 1, 2, 3, \ldots, 2011$. (Since the greatest common divisor of 2012 and 2015 is 1, the diagonal never passes through any point with integral coordinates—where the grid lines cross—that is interior to the rectangle.) Start with the unit square that has one vertex at (0,0), then go through another 2014 squares horizontally and 2011 squares vertically. The total number of such squares is $1+2014+2011 = \boxed{4026}$.

Contests written and compiled by Steven R. Conrad & Daniel Flegler

Problem 3-1

The diagonals of a rhombus are perpendicular bisectors of each other, so they split this rhombus into four 3-4-5 right triangles. The perimeter of the rhombus is $4 \times 5 = \boxed{20}$.

Problem 3-2

If the 2011th term is x, then $x+2 = 2x$, so $x = 2$. This implies that whenever any term is 2, every previous term is also 2. Consequently, every term is 2, and the sum of all 2012 terms is $\boxed{4024}$.

Problem 3-3

The sum of the squares of two real numbers is 0 only if both are 0. Solving $x-y-1 = 0$ and $x+y-39 = 0$, we get $(x,y) = \boxed{(20,19)}$.

Problem 3-4

Let the common difference be $d > 0$. Calling the middle term x, the 5 terms are $x-2d$, $x-d$, x, $x+d$, and $x+2d$. The sum is 540, so $5x = 540$, and $x = 108$. Since $x = 108$ and $d > 0$, the first term is $x-2d = 108-2d > 0$, so $d < 54$. To maximize the last term, let $d = 53$. In this case, $x+2d = \boxed{214}$.

Problem 3-5

The point whose coordinates we seek is the point at which the diagonals intersect. (A proof is in the note below.) The equations of the diagonals are $y-6 = -\frac{1}{9}(x+1)$ and $y = x+2$, and these lines intersect at the point with coordinates $\boxed{\left(\frac{7}{2},\frac{11}{2}\right)}$.

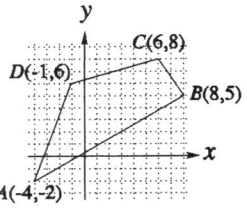

[NOTE: Draw the quadrilateral and its diagonals. Let them intersect at P. Pick any point Q other than P and draw $\overline{QA}$, $\overline{QB}$, $\overline{QC}$, and $\overline{QD}$. Since B, P, and D are collinear, $PB+PD < QB+QD$. Similary, use the other diagonal to prove that $PA+PC < QA+QC$. Adding the inequalities proves that the minimizing point is P, the intersection of the diagonals.]

Problem 3-6

The main difficulty is that the given inequality is not a strict one. It would be much easier to count the number of solutions if equality could be converted to inequality. Here's one way: We know that $1 \le a \le b \le c \le d \le 6$. If $A = a$, $B = b+1$, $C = c+2$, and $D = d+3$, it follows that $1 \le A < B < C < D \le 9$. Let's count. Every 4-tuple (A,B,C,D) gives rise to a solution $(A,B-1,C-2,D-3) = (a,b,c,d)$ whose coordinates satisfy the given inequality. The number of such ordered 4-tuples (A,B,C,D) is $\binom{9}{4} = 126$. Finally, the total possible number of ordered 4-tuples (a,b,c,d) is 6^4, so the probability is $\frac{126}{6^4} = \boxed{\frac{7}{72}} = 0.097\overline{2}$.

[NOTE: There are several ways to count the number of solution 4-tuples on a case-by-case basis.]

Contests written and compiled by Steven R. Conrad & Daniel Flegler   ©2012 by Mathematics Leagues Inc.

Problem 4-1

When a line crosses the y-axis, its x-coordinate is 0, so $a + 2013 = 0$ and $a = \boxed{-2013}$.

Problem 4-2

Method I: Since the length of a diagonal of the 12×16 rectangle is 20, the length of a diagonal of the smaller rectangle 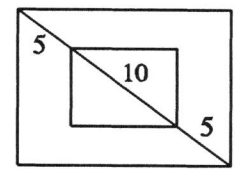 is $20 - (5+5) = 10$. By similar triangles, the smaller rectangle's sides are half as long as the larger rectangle's sides. Therefore, the area of the smaller rectangle is $\frac{12}{2} \times \frac{16}{2} = 6 \times 8 = \boxed{48}$.

Method II: Since the smaller rectangle's diagonals are half as long as those in the larger rectangle, and since these two rectangles are also similar, the area A of the smaller rectangle is $\left(\frac{1}{2}\right)^2 = \frac{1}{4}$ times the area of the larger rectangle, so $A = \frac{1}{4} \times 16 \times 12 = 48$.

Problem 4-3

Two days ago, x dogs skated. The number of dogs skating yesterday was 20% more than that, so it was $\frac{6x}{5}$. Today, we had 40% more dogs skating than yesterday. The number of dogs skating today (which must be an integer) is $\frac{7}{5}$ of $\frac{6x}{5} = \frac{42x}{25}$. This will be an integer whenever x is a multiple of 25. The least positive integral value of x is $\boxed{25}$.

Problem 4-4

This question really asks how many integers < 100 are rational powers of 8. Recall that $2 = 8^{1/3}$, and no other prime is a rational power of 8*. If we raise 8 to any non-negative multiple of 1/3, the result will be an integer. Since $x = 8^{k/3}$ is a positive integer < 100 if and only if $k = 0, 1, 2, 3, 4, 5,$ and 6, there is a total of $\boxed{7}$ values.

[**NOTE:** Here's a proof: Since x is a positive integer, it follows that $\log_8 x > 0$. If $\log_8 x = \frac{u}{v}$, where u and v are integers with $u \geq 0$ and $v > 0$, then $8^{u/v} = x$, or equivalently $2^{3u} = x^v$. By prime factorization, $x = 2^k$, where $k \geq 0$ is an integer. If $x = 2^k$, then $\log_8 x = \frac{k}{3}$, where $\frac{k}{3} \geq 0$ is rational number. Since $0 < x < 100$, the only solutions are $k = 0, 1, 2, 3, 4, 5,$ and 6.]

Problem 4-5

Draw the isosceles trapezoid and its circumcircle, as shown at the right. Draw a perpendicular from the center of the circle to both 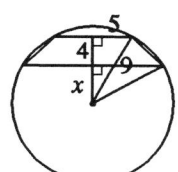 bases of the trapezoid. This bisects both of those bases, since a line through the center of a circle and perpendicular to a chord bisects the chord. Draw the radii shown in the diagram. Using the Pythagorean Theorem in both right triangles that have the radii as hypotenuses, we get $x^2 + 9^2 = r^2 = (x+4)^2 + 5^2$. Solving, $x = 5$. The area of the circle is $\pi r^2 = \boxed{106\pi}$.

Problem 4-6

Since $P(1) = P(2) = P(3) = P(4) = P(5) = 0$, we have $P(x) = k(x-1)(x-2)(x-3)(x-4)(x-5)$, where $k \neq 0$ is a real number. If we expand $P(x)$, the lead term will be kx^5. Since $kx^5 = x^5$, we know that $k = 1$. Now evaluate $P(-1)$: $P(x) = x^5 + ax^4 + bx^3 + cx^2 + dx + e$, so $P(-1) = (-1)^5 + a - b + c - d + e = -1(1 - a + b - c + d - e)$. From line 2 above, $P(-1) = (-2)(-3)(-4)(-5)(-6) = -720$, so $1 - a + b - c + d - e = \boxed{720}$.

Contests written and compiled by Steven R. Conrad & Daniel Flegler   ©2013 by Mathematics Leagues Inc.

Problem 5-1

Since both right triangles seen at the right are $30°$-$60°$-$90°$ triangles, the lengths of the labeled sides are in the ratio 2:1. Since the x-coordinate of the rightmost vertex of the hexagon is 4, the lengths shown in the diagram are accurate. The regular hexagon's perimeter is $\boxed{12}$.

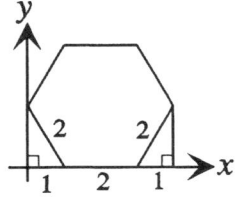

Problem 5-2

We know that $x^2 - 100x + 2500 = (x-50)^2 \geq 0$ is always positive unless $(x-50)^2 = 0$. The only integer x that satisfies $(x-50)^2 = 0$ is $x = \boxed{50}$.

Problem 5-3

The most important thing to realize is that if an integer becomes a perfect square after being doubled, it was twice a perfect square before being doubled. If our number is N, we can write $N = 2k^2$, where $k > 0$ is an integer. Since $N = 2k^2 < 1000$, $k^2 < 500$, so $k < 22.36\ldots$. The largest integral value of k is 22, so the largest possible N is $2 \times 22^2 = 2 \times 484 = \boxed{968}$.

Problem 5-4

It is clear that $0 < \cos^2 x < 1$, or else the sum of all the terms would be either 0 or not finite. In the infinite geometric series $a + ar + ar^2 + ar^3 + \ldots$ with common ratio r, if $0 < r < 1$, the sum S is given by $S = \frac{a}{1-r}$. In the cosine series, $a = \cos^2 x$ and $r = \cos^2 x$. Since $0 < r = \cos^2 x < 1$, S is $\frac{\cos^2 x}{1-\cos^2 x} = 2013$. Solving, we get $\cos^2 x = \frac{2013}{2014}$ and $\sin^2 x = 1 - \cos^2 x = \frac{1}{2014}$. The sine series sum is $\frac{\sin^2 x}{1-\sin^2 x} = \frac{1}{2014} \div \frac{2013}{2014} = \boxed{\frac{1}{2013}}$.

Problem 5-5

Method I: If r is the number of red cookies, and b is the number of blue cookies, then $r+b = 100$. The number of ways I can choose a pair of red cookies is $\binom{r}{2} = \frac{r(r-1)}{2}$. The number of ways I can choose a pair of blue cookies is $\binom{b}{2} = \frac{b(b-1)}{2}$. I can choose 1 of each color in rb ways, so $\frac{r(r-1)}{2} + \frac{b(b-1)}{2} = rb$. Clearing fractions and rearranging, $r^2 - 2rb + b^2 = r+b$. Since $r+b = 100$, $r^2 - 2rb + b^2 = (r-b)^2 = 100$, and $r-b = \pm 10$. Since $r+b = 100$, $r = 45$ or $r = 55$. Therefore, at the very least, r is $\boxed{45}$.

Method II: The number of ways I can choose 2 cookies of the same color equals the number of ways I can choose a pair of cookies with 1 cookie of each color, so the total number of ways I can choose 2 of the 100 cookies in any color combination is twice the number of ways I can choose a pair of cookies with 1 cookie of each color. Thus, $\binom{100}{2} = 2 \times \binom{r}{1} \times \binom{100-r}{1}$, so $50 \times 99 = 2r(100-r)$. Hence $r^2 - 100r + 2475 = (r-45)(r-55) = 0$. At the very least, $r = 45$.

Problem 5-6

We are told that $N = ab - (a+b)$ is a positive integer. Since a and b are positive integers, solve for each to get $a = \frac{N+b}{b-1} = 1 + \frac{N+1}{b-1}$ and $b = 1 + \frac{N+1}{a-1}$. For a and b to be positive integers, both $\frac{N+1}{b-1}$ and $\frac{N+1}{a-1}$ must be positive integers, implying that both $b-1$ and $a-1$ are factors of $N+1$. That means that there is only one pair of positive integers (a,b), with $a \leq b$, for which $N = ab - (a+b)$ if and only if the numerator, $N+1$, is a prime. (The argument is reversible.) The number of values of N, with $40 < N < 50$, for which $N+1$ is prime is the number of primes $N+1$ that are both greater than 41 and less than 51. Thus $N+1 = 43$ or 47. The required values of N are $\boxed{42, 46}$.

Contests written and compiled by Steven R. Conrad & Daniel Flegler   ©2013 by Mathematics Leagues Inc.

Problem 6-1

If each side of square A has length s, then each side of square B has length $s-1$. The areas of the squares differ by 2013, so $2013 = s^2 - (s-1)^2 = 2s-1$. Solving, $s = 1007$ and the perimeter $= 4s = \boxed{4028}$.

Problem 6-2

Since the ordered pair given in the question corresponds to $(1)(94)$, we are looking for the ordered pair corresponding to $94 = 2 \times 47 = (x-y)(x+y-1)$. Since $x+y-1 \geq x > x-y$, the smaller factor is $2 = x-y$ and the larger factor is $47 = x+y-1$. Adding, $2x-1 = 49$, so $x = 25$. The other pair is $\boxed{(25,23)}$.

Problem 6-3

If every player tied every game, all players would be tied with a total of 3.5. Thus a winning total is greater than or equal to 3.5. A single winner's total must be greater than average. Can a single winner have a total of 4? Yes. If every game except one is tied, then 5 of the players have a total of $7 \times 0.5 = 3.5$, one player has a 3, and one has a winning total of $\boxed{4}$.

Problem 6-4

Each of the points labeled with a C could serve as the third vertex of isosceles $\triangle ABC$. The total number of such points is $\boxed{5}$.

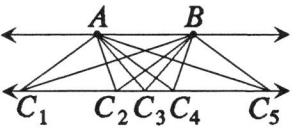

[**NOTE:** Using center A and radius AB, we get points C_1 and C_4. Using center B and radius AB, we get points C_2 and C_5. Finally, the perpendicular bisector of $\overline{AB}$ produces point C_3.]

Problem 6-5

Let $\frac{21}{x} + \frac{70}{y} = n$, where n is an integer. Clearing fractions and rearranging, $21y = x(yn-70)$. Since x divides the right side, it also divides the left side, $21y$. Since $\frac{21}{x}$ is already in lowest terms, the GCD of x and 21 is 1. Therefore, x divides y. To show that y divides x, clear fractions and rearrange terms in $\frac{21}{x} + \frac{70}{y} = n$ to get $70x = y(xn-21)$. In a manner similar to what was done above, we can show that y divides x. If x divides y and y divides x, then $x = y$. Add $\frac{21}{x}$ and $\frac{70}{y}$ to get $\frac{91}{x} = n$, an integer. The only possible values of x are 1, 7, 13, and 91. Both $\frac{70}{y} = \frac{70}{x}$ and $\frac{21}{x}$ are already in lowest terms, so $x = 1$ or 13. Thus, the second solution must be the ordered pair $\boxed{(13,13)}$.

Problem 6-6

Method I: From the center of the circle, draw perpendiculars to each chord, bisecting each chord. When two chords intersect, the product of the lengths of the segments of one chord equals the product of the lengths of the segments of the other chord. Thus, one of the chords has segments with lengths 4 and 18. The other has segments with lengths 6 and 12. The lengths of the chords are $4+18 = 22$ and $6+12 = 18$. The lengths of the various segments are as marked in the diagram. By the Pythagorean Theorem, in the right $\triangle$, $r^2 = 7^2 + 9^2 = 130$, so the area of the circle is $\boxed{130\pi}$.

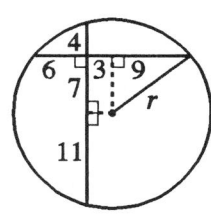

Method II: It is a little-known fact that if the four segments have lengths a, b, c, and d, then the length of the circle's diameter is $\sqrt{a^2 + b^2 + c^2 + d^2}$. We'll print the best submitted proof of this formula in the final contest newsletter. We'll also print the solution of an outstanding mathematician, written when he was a high school student more than 50 years ago.

Problem 1-1

The perimeter of each rectangle is 18, so $6x = 18$, and $x = 3$. The dimensions of each rectangle are 3 and 6, so the area of each is $3 \times 6 = \boxed{18}$.

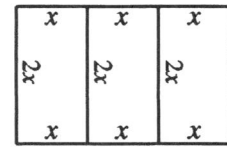

Problem 1-2

The greatest common factor of the two chosen numbers is 24, so each must be a multiple of 24. The largest multiples of 24 that are less than 100 are $3 \times 24 = 72$ and $4 \times 24 = 96$. Their greatest common divisor is 24, and their sum is $3 \times 24 + 4 \times 24 = 7 \times 24 = \boxed{168}$.

Problem 1-3

The number $BBB/3 = ABC$ is a 3-digit number that has BBB's middle digit, B, as its own middle digit. Of $BBB = 111, 222, 333, 444, 555, 666, 777, 888,$ and 999, division by 3 leaves the middle digit unchanged only for $444/3 = 148$, so $(A,B,C) = \boxed{(1,4,8)}$.

Problem 1-4

The number 10^{2013} has 2014 digits, 2013 of which are 0s. The result of the subtraction $10^{2013} - 2013$ is $1000\ldots00 - 2013 = 999\ldots997\,987$, a number with 2013 total digits. The sum of the last 4 digits is $7+9+8+7 = 31$. The first $2013 - 4 = 2009$ digits have a sum of $2009 \times 9 = 18\,081$. The sum of all 2013 digits is $31 + 18\,081 = \boxed{18\,112}$.

Problem 1-5

In the diagram, $a^2 + 4b^2 = 4^2$ and $4a^2 + b^2 = 3^2$. Adding, we get $5a^2 + 5b^2 = 16 + 9 = 25$. Dividing both sides by 5 and then multiplying both sides by 4, we get $4a^2 + 4b^2 = 20 = \text{hypotenuse}^2$, so hypotenuse's length is $\boxed{\sqrt{20}}$.

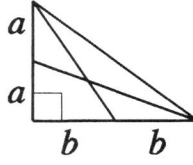

Problem 1-6

If 100 000 surfboards were made, 100 would be bad and 99 900 would be good. Of the 100 bad surfboards, 99 would correctly test bad. Of the 99 900 good surfboards, 1% or 999 would test bad. There are 1098 surfboards that would test bad, and 99 of these would actually be bad. So the probabilty of a board being bad that tested bad is 99/1098 or about $\boxed{9\%}$.

Contests written and compiled by Steven R. Conrad & Daniel Flegler   ©2013 by Mathematics Leagues Inc.

Problem 2-1

In any set of three consecutive integers, the middle integer is one-third the sum of the three integers. Since that sum must be divisible by 3, the required sum must be a perfect square greater than 9 (since $2+3+4 = 9$) that is divisible by 9. The least such sum is 36. The consecutive integers are 11, 12, and 13, and the date is $\boxed{11\text{-}12\text{-}13}$.

Problem 2-2

Since $a^2-b^2 = b-a$, we have $(a+b)(a-b) = -1(a-b)$. Since $a \neq b$, $(a-b) \neq 0$. Dividing the second equation through by $(a-b)$, we get $a+b = \boxed{-1}$.

Problem 2-3

Since $2^{67}+2^{71} = 2^{67}(1+2^4) = 2^{67}(17)$, the only odd prime factor of $2^{67}+2^{71}$ is $\boxed{17}$.

Problem 2-4

The volume of the cube is $6^3 = 216$. From the cube, a right pyramid was removed. The pyramid's base is a 3-4-5 right triangle, and the height of the pyramid is 5. The pyramid's base has area $bh/2 = (3 \times 4)/2 = 6$. The volume is $Bh/3$, where B is the area of the base, so the volume of the pyramid is $(6 \times 5)/3 = 10$. The volume of the remaining solid is $6^3-10 = 216-10 = \boxed{206}$.

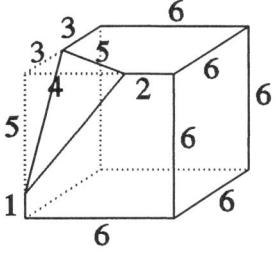

Problem 2-5

Can the number of 1's be 100? No, the product is 1, but the sum is 100. Can the number of 1's be 99? No, the product is the value of the other number, and the sum is 99 more than that. Can the number of 1's be 98? Yes, since the sum of 98 1's, a 2, and a 100 is $98+2+100 = 200 = 1 \times 1 \times \ldots \times 1 \times 2 \times 100 =$ the product. The maximum number of 1's is $\boxed{98}$.

[**NOTE:** To find all solutions with 98 1's, we can solve $98+x+y = xy$, or $98+x = y(x-1)$, so $y = (98+x)/(x-1) = (x-1 + 99)/(x-1) = 1 + 99/(x-1)$. Now, set $x-1$ equal, in turn, to each divisor of 99.]

Problem 2-6

Since the diagonals of the quadrilateral are congruent, the upper two overlapping isosceles triangles are congruent to each other by *SSS*, so the four base angles, each marked with an x, are congruent. The lower two overlapping isosceles triangles are congruent to each other by *SSS*, so their vertex angles, each marked with a v, are congruent. Look at the upper triangle with an angle marked with an a. Its other two angles are congruent, since each is marked with an x. In the lower such triangle with an angle marked with an a, the other two angles are also congruent, so those two angles must both also be marked with an x, just like the upper such triangle. Finally, the base angles of the lower overlapping isosceles triangles are congruent. Since $v = x$, each base angle $b = x+v = 2x$. From the diagram, $10x = 360°$. Solving, $x = 36°$ and $3x = 108°$ or $\boxed{108}$.

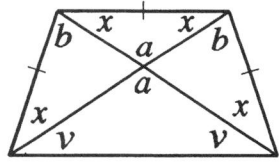

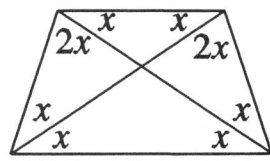

NOTE: Connect 4 vertices of any regular pentagon (interior angle 108°), and you'll get this quadrilateral.

Contests written and compiled by Steven R. Conrad & Daniel Flegler   © 2013 by Mathematics Leagues Inc.

Problem 3-1

Since the area of the circle is 16π, the length of each radius of the circle is 4. Since the triangle is equilateral, its perimeter is $3 \times 4 = \boxed{12}$.

Problem 3-2

My uncle's age must be a two-digit number, so we can represent his age as $10t+u$, where t and u are one-digit numbers. We know that $10t+u = 6(t+u)$, from which we get $4t = 5u$. The only solution that meets the conditions of the problem is $(t,u) = (5,4)$, so my uncle's age is $\boxed{54}$.

Problem 3-3

The right-most entry in row n is the number n^2, so the final entry in row 2012 is 2012^2. Therefore, the 2013th entry in the next row, the 2013th row, is $2012^2 + 2013 = \boxed{4\,050\,157}$.

Problem 3-4

Consider the function $P(x) = (x+1)(x+4)(x+9)$, whose roots are the real numbers -1, -4, and -9. Since $P(x^2) = (x^2+1)(x^2+4)(x^2+9) > 0$ whenever x is real, and since $P(x^2) = 0$ has only imaginary roots, there is at least one such polynomial equation for which the number of real roots is $\boxed{0}$.

Problem 3-5

Method I: In our 183-term sequence $a, a+d, a+2d, a+3d, \ldots, a+182d$, the 72nd term is $a+71d$, the 112th term is $a+111d$, and the middle term is $a+91d$. We're told that the sum of the 72nd term and 112th terms is 22, so $22 = (a+71d) + (a+111d) = 2a+182d = 2(a+91d)$. Thus, $a+91d = 11$ is the value of the middle term = the value of the average term, and the sum of all 183 terms is $183 \times 11 = \boxed{2013}$.

Method II: $a + (a+d) + (a+2d) + \ldots + (a+182d) = 183a + \frac{183}{2}(182d) = 183(a+91d)$. As in Method I, $a+91d = 11$, so the sum is $183 \times 11 = 2013$.

Problem 3-6

Each of the 5 integers is paired with each of the other 4 integers, so each integer appears in 4 of the 10 sums. If we add all 10 sums together, their sum is 84. That means that the sum of the five integers is $84/4 = 21$. The smallest sum is 1, so the sum of the two smallest integers is 1. The largest sum is 15, so the sum of the two largest integers is 15. Clearly then, the middle integer is $21-15-1 = 5$. The second smallest sum has to be the sum of the smallest and middle integers, so the smallest integer is $4-5 = -1$. Similarly, the second largest sum, 14, is the sum of the largest and middle integers, so the largest integer is $14-5 = 9$. To get the second smallest integer, subtract the smallest integer from the sum of the two smallest integers; so the second smallest integer is $1-(-1) = 2$. Since the largest integer is 9 and the sum of the two largest integers is 15, the second largest integer is 6. The 5 integers are $\boxed{-1, 2, 5, 6, 9}$.

Contests written and compiled by Steven R. Conrad & Daniel Flegler   ©2013 by Mathematics Leagues Inc.

Problem 4-1

We are told that, for every number on first list (call it x), there is one number on the other list (call it y) so that the product of the two numbers stays the same. Therefore, the largest number on one list must be paired with the smallest number from the other list, so $xy = 2 \times 180 = 360$. The number that gets paired with 30 is the value of x for which $(x)(30) = 360$. That value is $\boxed{12}$.

[**NOTE:** This is called an inverse variation.]

Problem 4-2

We are told that at least one root is an integer. The sum of the roots of the first equation is a, and the sum of the roots of the second equation is b. Since both a and b are integers, the other root of each equation is also an integer. The product of the roots of the first equation is 2014. The product of the roots of the second equation is 2015. The largest common integral factor of any two consecutive integers is $\boxed{1}$.

Problem 4-3

When First had gone 10 km, Second had gone 8 km and Third had gone 6 km. Since the ratio of the distances traveled by Second and Third is 8:6 = 4:3, Third traveled 3/4 of the distance traveled by Second. When Second finished the 10 km race, Third had gone 7.5 km; so Second won by $\boxed{2.5}$ km.

Problem 4-4

The circle's area is π, so its radius is 1. There is a theorem in mathematics that says that of all triangles that can be inscribed in a given circle, the equilateral triangle has the greatest area. To make the *longest* side of an inscribed triangle as *small* as possible, all sides must have the same length. Here's why: Any triangle of area A whose base is shorter than the base of an equilateral triangle of area A must have legs that are longer than those of the equilateral triangle. The length of one side of the equilateral triangle that is inscribed in a circle of radius 1 is $\boxed{\sqrt{3}}$.

Problem 4-5

$n + \ldots + (n+98) = \frac{99}{2}\big(n + (n+98)\big) = \frac{99}{2}(2n+98) = 99(n+49) = (3^2 \times 11)(n+49) = P$. The least positive perfect-cube value of P is $3^3 \times 11^3$. To find the n for this value of P, set $(n+49) = 3 \times 11^2 = 363$. Solving, $n = 363 - 49 = \boxed{314}$.

Problem 4-6

The shaded region **below** the line through (0,0) and (4,1) consists of a square (labeled 1 in the diagram) plus the lower half of a 1×4 rectangle with (0,0) and 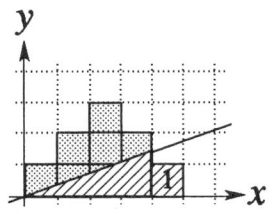 (4,1) as opposite vertices. The total area below that line is 3, so the line we want must cross $x = 4$ above (4,1). We need a total shaded area of 9/2 = 4.5. If the line through the origin crosses $x = 4$ at (4,y), then $1 + bh/2 = 1 + 4y/2 = 4.5$, so $y = 7/4$. The slope of the line that connects (0,0) to (4,7/4) is $\boxed{\frac{7}{16}}$.

Contests written and compiled by Steven R. Conrad & Daniel Flegler   ©2014 by Mathematics Leagues Inc.

Problem 5-1

Factoring, $(x-1)(x-2) = 0$, so $x^2-3x+2 = 0$ is satisfied when $x = 1$ or 2. Similarly, $(x+1)(x-2) = 0$, so $x^2-x-2 = 0$ is satisfied when $x = -1$ or 2. The value of x that satisfies the first equation but not the second is $x = \boxed{1}$.

Problem 5-2

The area of the square is the same as the area of the circle. Clearly, their overlapping regions are the same, so the area of their common region is the same. Subtracting, their non-overlapping regions must have the same total area. The difference is therefore $\boxed{0}$.

Problem 5-3

Since the sum of the measures of the interior angles of a convex quadrilateral is 360, a convex quadrilateral can have neither 4 acute angles nor 4 obtuse angles. Since a convex quadrilateral can have three 80° angles and one 120° angle, $m = 3$. Since a convex quadrilateral can have three 95° angles and one 75° angle, $M = 3$. Therefore, $m+M = 3+3 = \boxed{6}$.

Problem 5-4

Work backwards. At the end, "8 more than half the remaining clothes pins are blue," so after counting half, 8 remained (and we're told that they were all blue). Since half the number of clothes pins remaining is 8, the number of blue clothes pins—the only color remaining—was 16. To get the number of green clothes pins, consider the clothes pins remaining after removing the red ones. This mix of green and blue clothes pins was first divided in half, and 1 less than half were green; so 1 more than half, 16, were blue. Therefore, 1 less than half, 14, were green. After removing the red clothes pins, 30 clothes pins remained. We're told that 1 more than half the clothes pins were red. Hence, the number of non-red clothes pins = 30 = 1 less than half. Finally, the number of red clothes pins was 32, and the garbage manager has a total of $\boxed{62}$ clothes pins.

Problem 5-5

Whenever z is a root of the polynomial equation, so is iz. Hence, since 2 is a root, $2i$ is a root. Since $2i$ is a root, $(2i)i = -2$ is a root . . . , from we see that the product of 2 and any positive integral power of i is a root. The powers of i are $\{1, i, -1, -i\}$, so the only 4 numbers that MUST be roots are $\boxed{2,\ 2i,\ -2,\ -2i}$.

Problem 5-6

An easy way to cube a sum is to use the identity, easily verified, that $(u+v)^3 = u^3+v^3+3uv(u+v)$. Let $u = \sqrt[3]{x+\sqrt{x^2+a^3}}$ and $v = \sqrt[3]{x-\sqrt{x^2+a^3}}$. We're told that $u+v = a$. Using that sum, if we now cube both sides and then combine like terms, we'll get $2x + 3\left(\sqrt[3]{x^2-x^2-a^3}\right)(a) = a^3$, so $2x - 3a^2 = a^3$. Solving for x, we get $x = \boxed{\dfrac{a^3 + 3a^2}{2}}$.

Contests written and compiled by Steven R. Conrad & Daniel Flegler   ©2014 by Mathematics Leagues Inc.

Problem 6-1

Since $x^2-x = 2^2-2 = 2$, $x^2-x-2 = (x-2)(x+1) = 0$; so $x = \boxed{2, -1}$.

Problem 6-2

In the diagram at the right, the circle with radius 6 has been removed, and the 3 externally tangent circles are shown. As you can see in the diagram, the length

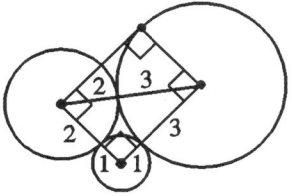

of each side of the rectangle is the sum of the lengths of the radii of two of the circles. The perimeter of the rectangle is $2(3+4) = \boxed{14}$.

NOTE: Proving that the center of the circle to which the other three are internally tangent has its center on the circle with radius 3 is very interesting. Go to http://www.ams.org/samplings/feature-column/fcarc-kissing for an excellent American Mathematical Society discussion of "kissing circles."

Problem 6-3

The first position can be occupied by any of 6 people. The second position must be filled by one of the 3 people of a different gender than the first person, and so on. The total number of different ways to line up these six people is $6 \times 3 \times 2 \times 2 \times 1 \times 1 = 2(3!)^2 = 2(6^2) = \boxed{72}$.

Problem 6-4

The longest of the three given lengths must be the hypotenuse, so the side with length $\sqrt{\log 3}$ cannot be the hypotenuse. Applying the Pythagorean Theorem, either $\log x + \log 3 = \log 4$ (so $3x = 4$, or $x = 4/3$), or $\log 3 + \log 4 = \log x$ (so $12 = x$). The two values of x are $\boxed{4/3, 12}$.

Problem 6-5

Since ℓ and m are perpendicular, (slope of ℓ)(slope of m) $= -1$. Since the product of all three slopes is -8, the slope of n must be 8. Since ℓ and n are perpendicular, the product of their slopes is -1, from which we can conclude that the slope of $\ell = \boxed{-1/8}$.

Problem 6-6

$2^2 < \sqrt[4]{2014} = x^{[x]} \approx 6.7 < 3^2$, so $2 < x < 3$. Thus, $[x] = 2$, so the equation we're solving is $x^2 = \sqrt[4]{2014}$. Solving, $x = \boxed{(2014)^{(1/8)}} = \sqrt[8]{2014}$.

Contests written and compiled by Steven R. Conrad & Daniel Flegler   ©2014 by Mathematics Leagues Inc.

Problem 1-1

Set as equal the lengths of any two of the rectangle's sides. No matter which two lengths are set equal, the value of x is 1 , so the area is $1 \times 1 = \boxed{1}$.

Problem 1-2

Since $2\,326\,045$ is divisible by 5, the other prime is $\frac{2\,326\,045}{5} = \boxed{465\,209}$.

Problem 1-3

Solving for 2014^n and factoring out 2014^{2012}, we get

$$2014^n = 2014^{2012}(2014^2 - 2013 \times 2014 - 2013)$$
$$= 2014^{2012}(2014^2 - 2013(2014 + 1))$$
$$= 2014^{2012}(2014^2 - 2013 \times 2015)$$
$$= 2014^{2012}(1) = 2014^{2012}, \text{ so}$$
$$n = \boxed{2012}.$$

Problem 1-4

Join 2 adjacent vertices of the star, say A and E. In $\triangle AEM$ and $\triangle BDM$, since the angles at M are congruent, the sum of angles B and D of $\triangle DBM$ is equal to the sum of angles A and E of $\triangle AEM$. There-fore, the sum of the angles at the vertices of the star is the same as the sum of the angles of $\triangle ACE$, and the sum of those degree-measures is $\boxed{180}$ or 180°.

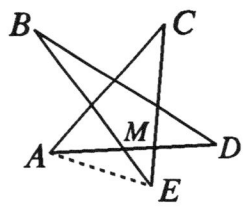

[**NOTE:** The problem implies that the result is independent of the shape of the pentagon. Using a regular pentagon (whose interior angles are each 108°), the answer is easily seen to be $5 \times 36° = 180°$.]

Problem 1-5

$2^{\frac{1}{2}} \times 3^{\frac{1}{3}} \times 4^{\frac{1}{4}} = 2^{\frac{1}{2}} \times 3^{\frac{1}{3}} \times \left(4^{\frac{1}{2}}\right)^{\frac{1}{2}} = 2^{\frac{1}{2}} \times 3^{\frac{1}{3}} \times \left(2\right)^{\frac{1}{2}} = 2 \times 3^{\frac{1}{3}} = 8^{\frac{1}{3}} \times 3^{\frac{1}{3}} = 24^{\frac{1}{3}}$, so $(k,b) = \boxed{(3,24)}$.

Problem 1-6

Method I: Arrange the original stack of 8 cards in a circle. The first 4 cards chosen are the 1st (A), 3rd (K), 5th (A), and 7th (K) cards, leaving the 2nd, 4th, 6th, and 8th cards. Continuing "skip, then choose," with the still-unchosen cards, we skip the 8th, choose the 2nd (A), skip the 4th, choose the 6th (K). Finally, skip the 8th and choose the 4th (A), leaving the 8th (K). The result, in order, is $\boxed{AAKAAKKK}$.

Method II: After numbering the cards 1, 2, 3, 4, 5, 6, 7, 8, remove the cards in order:

First Draw: 1 is removed and 2 is moved to the bottom. The remaining cards, in order, are 3, 4, 5, 6, 7, 8, 2.

Second Draw: 3 is removed and 4 is moved to the bottom. The remaining cards, in order, are 5, 6, 7, 8, 2, 4.

Third Draw: 5 is removed, and 6 is moved to the bottom. The remaining cards, in order, are 7, 8, 2, 4, 6.

Fourth Draw: 7 is removed, and 8 is moved to the bottom. The remaining cards, in order, are 2, 4, 6, 8.

Fifth Draw: 2 is removed, and 4 is moved to the bottom. The remaining cards, in order, are 6, 8, 4.

Sixth Draw: 6 is removed and 8 is moved to the bottom. The remaining cards, in order, are 4, 8.

Seventh Draw: 4 is removed, and 8 is left over. The remaining card is 8.

The cards were removed in the order **1**, 3, **5**, 7, **2**, 6, **4**, 8, so cards **1**, **5**, **2**, and **4** must be **Aces**. The other four cards must be *Kings*. Now return the **Ace** and *King* cards to their original numerical order. In order, the ranks of the cards are $AAKAAKKK$.

Method III: The 8th-revealed card was a K, and the 7th-revealed card that was on top of it was an A. Now reverse the steps from the given procedure: move the card now on the bottom of the 2-card stack (the K) to the top, then put the 6th-revealed card (a K) back to the top of the stack. Repeat the procedure, moving the card now on the bottom of the stack to the top, and adding back to the stack on top of that the latest-revealed card not yet returned to the stack. Continue these steps till all 8 cards are back in the stack.

Contests written and compiled by Steven R. Conrad, Daniel Flegler, & Adam Raichel © 2014 by Mathematics Leagues Inc.

Problem 2-1

The larger circle's area is $16\pi = \pi R^2$, so the larger circle's radius is $R = 4 =$ the diameter of the smaller circle. Therefore, the radius of the smaller circle is $r = 2$, and that circle's area is $\pi r^2 = \boxed{4\pi}$.

Problem 2-2

Factoring the numerator of the right side, we get that $2015!-2014! = 2014!(2015-1) = 2014!(2014)$. Divide this by 2014, the denominator of the right side, to get $2014!$. Therefore, $n! = 2014!$, so $n = \boxed{2014}$.

Problem 2-3

Method I: When I divide x by y, I get a quotient of 3, so $x > y > 0$. Therefore, when I divide y by x, the quotient is 0 and the remainder equals y. Therefore, $y = 12$ and $x = 3(12) + 7 = \boxed{43}$.

Method II: Since $x = 3y+7$ and $y = kx+12$, we can substitute to get $x = 3(kx+12)+7 = 3kx+43$, so $x(1-3k) = 43$. Since k is a non-negative integer, the only solution occurs when $k = 0$, from which $x = 43$.

Problem 2-4

Let the roots be r and $2r$. The product of the roots $= 2r^2 = a^2+2$, so $18r^2 = 9a^2+18$. The sum of the roots is $3r = 2a+1$, so $2(3r)^2 = 2(2a+1)^2$, or $18r^2 = 8a^2+8a+2$. Setting $9a^2+18 = 8a^2+8a+2$ and solving, $a^2-8a+16 = (a-4)^2 = 0$, so $a = \boxed{4}$.

Problem 2-5

Method I: If I began with $\$x$, I'd have $\$(x-32)$ after my first $\$32$ withdrawal. I double to $\$(2x-64)$. I have $\$(2x-96)$ after taking $\$32$. I double to $\$(4x-192)$. I then have $\$(4x-224)$ after taking $\$32$. I double to $\$(8x-448)$ and have $\$(8x-480)$ after taking $\$32$. For the final round, I have $\$(16x-960)$ and $\$(16x-992)$ after taking $\$32$. The account was then worthless, so we can write $16x-992 = 0$, so the number of dollars with which I opened the account was $\boxed{62}$ or $\$62$.

Method II: Start at the end and work **backwards**, writing all five of the ordered pairs that have the form ($ after $32 taken, $ before $32 taken). The five such pairs are ($0, $32), ($16, $48), ($24, $56), ($28, $60), and ($30, $62).

Problem 2-6

Method I: Imagine that we are going to place 20 balls in a straight line to represent the first 20 positive integers. The 6 integers we will choose will be represented by black balls that must be stuck in among the 14 we don't choose, represented by red balls. Each time a black ball is placed, it must be in one of the 15 spots: the 13 that are between red balls, and the 2 that are at the ends of the straight line. Each possible positioning of the 6 black balls into these 15 spots corresponds to choosing 6 of the first 20 positive integers with no two consecutive. There are $\binom{15}{6} = \boxed{5005}$ ways to choose these 6 integers with no two consecutive.

Method II: Let a, b, c, d, e, f be 6 integers (no two consecutive) that are chosen at random from $S = \{1, 2, 3, \ldots, 20\}$, with $a < b < c < d < e < f$. Create a new set of numbers $A = a, B = b-1, C = c-2, D = d-3, E = e-4, F = f-5$, so that A, B, C, D, E, F are 6 of the first 15 positive integers. There is a 1-1 correspondence between each selection of 6 integers (no two consecutive) from set S and the corresponding set $\{A, B, C, D, E, F\}$ from $\{1, 2, 3, \ldots, 15\}$. Therefore, there are $\binom{15}{6} = 5005$ ways to choose 6 of the first 20 positive integers such that no two are consecutive.

Contests written and compiled by Steven R. Conrad, Daniel Flegler, & Adam Raichel   ©2014 by Mathematics Leagues Inc.

Problem 3-1

Method I: If each small rectangle has dimensions ℓ and w, then the large rectangle has dimensions 2ℓ and $2w$, so the large rectangle's perimeter is twice that of the small rectangle = $2(2014)$ = $\boxed{4028}$.

Method II: If one side of a a small rectangle has length x, then the other side has length $1007-x$. The sides of the large rectangle have lengths $2x$ and $2014-2x$, so the large rectangle's perimeter is $2[2x+(2014-2x)] = 4028$.

	$1007\text{-}x$	$1007\text{-}x$
x		
x		

Problem 3-2

Cubing, we get $(x+y)^3 = x^3 + 3x^2y + 3xy^2 + y^3 = (x^3+y^3) + 3xy(x+y) = 200 + 300 = \boxed{500}$.

Problem 3-3

Method I: The rates had the ratio $45{:}30 = \frac{3}{2}$, so the times taken had the ratio $\frac{2}{3}$. Let the time Duck flew be t hours. Hawk began 2 hours after Duck, but arrived 1 hour earlier, so Hawk flew for $t-3$ hours. Hence, $\frac{2}{3} = \frac{t-3}{t}$, and $t = 9$. In km, the distance from A to B is $9 \times 30 = \boxed{270}$ or 270 km.

Method II: Using the notation above, $30t = 45(t-3)$, so $t = 9$. The distance traveled was 270 km.

Problem 3-4

Method I: Since $f(x) = 2^x f(1-x), f(1-x) = 2^{1-x}f(x)$. Subsituting, we get $f(x) = (2^x)(2^{1-x})[f(x)] = 2f(x)$. Therefore, $f(x) = 0$ for all x, so $f(3) = \boxed{0}$.

Method II: If $x = 3$, then $f(3) = 8f(-2)$. If $x = -2$, then $f(-2) = \frac{1}{4}f(3)$, so $f(3) = 8(\frac{1}{4})f(3) = 2f(3)$. Finally, $f(3) = 2f(3)$ if and only if $f(3) = 0$.

Problem 3-5

In a right triangle the square of the longest side equals the sum of the squares of the other 2 sides. In an acute triangle, the square of the longest side is less than the sum of the squares of the other 2 sides. The triangle's longest side can't be x. If the longest side is 5, then $5^2 < x^2 + (x+1)^2 \Leftrightarrow x^2+x-12 = (x+4)(x-3) > 0$. The solutions are $\{x \mid 3 < x < 5\}$. If the longest side is $x+1$, then $(x+1)^2 < 5^2+x^2$, whose solutions are $\{x \mid 4 \le x < 12\}$. The values of x for which either condition are met are the values of x for which $\boxed{3<x<12}$.

Problem 3-6

If a and b are a pair of positive integers whose least common multiple = $\mathrm{lcm}(a,b) = 540 = (2^2)(3^3)(5^1)$, then a and b contain only powers of 2, 3, and 5. Let's let $a = (2^x)(3^y)(5^z)$ and $b = (2^r)(3^s)(5^t)$.

If $x = 0$ or 1, then $r = 2$ since lcm (a,b) has a factor of 2^2. If $x = 2$, then $r = 0, 1,$ or 2 since lcm (a,b) has a factor of 2^2. In all there are 5 possible pairs of 2^x and 2^r such that lcm (a,b) has a factor of 2^2.

If $y = 0, 1, 2$, then $s = 3$ since lcm (a,b) has a factor of 3^3. If $y = 3$, then $s = 0, 1, 2, 3$ since lcm (a,b) has a factor of 3^3. There are 7 possible pairs of 3^y and 3^s such that lcm (a,b) has 3^3 as a factor.

If $z = 0$, then $t = 1$ since lcm (a,b) contains 5^1. If $z = 1$, then $t = 0$ or 1 since lcm (a,b) contains 5^1. In all there are 3 possible pairs of 5^z and 5^t such that lcm (a,b) contains 5^1. Multiplying the total number of these values of a and b, we get $5 \times 7 \times 3 = 105$.

Excepting $a = b = (2^2)(3^3)(5^1)$, we counted each possibility twice, such as $a = (2^1)(3^1)(5^1), b = (2^2)(3^3)(5^1)$ and $a = (2^2)(3^3)(5^1), b = (2^1)(3^1)(5^1)$.

Let's remove duplicates from the list. The 104 twice-listed possibilities are 52 possibilities after removal of duplicates. When we include the one unduplicated value, $(2^2)(3^3)(5^1)$, the answer is $52+1 = \boxed{53}$.

Contests written and compiled by Steven R. Conrad, Daniel Flegler, & Adam Raichel   ©2014 by Mathematics Leagues Inc.

Problem 4-1

Since $\frac{x - 2015 + x + 2015}{2} = 2015$, $x = \boxed{2015}$.

Problem 4-2

Two sides of rt. $\triangle$I are congruent to two sides of rt. $\triangle$II, but $\triangle$I $\not\cong$ $\triangle$II. To make $\triangle$II the smaller triangle, choose to make the shortest sides of $\triangle$I (3 and 4) the sides that are congruent to sides of $\triangle$II. To keep the triangle as small as possible, make the hypotenuse (as opposed to a leg) of $\triangle$II (the smaller triangle) have length $\boxed{4}$.

[The smaller triangle's legs have lengths 3 and $\sqrt{7}$.]

Problem 4-3

Factoring, $5x^{301} + 4x^{300} = x^{300}(5x+4)$. Since x^{300} is a perfect cube, this equation's right side will be a perfect cube if and only if $5x+4$ is a perfect cube. Let's solve $5x+4 =$ perfect cube (starting with $1^3 = 1$) by trial and error until we get an integral solution. The first such integral solution arises from $5x+4 = 4^3 = 64$, when $x = \boxed{12}$.

Problem 4-4

If you were told that there is some integer whose square is 324, how would you find the square of the next consecutive integer? You would evaluate that $\sqrt{324} = 18$, then add 1 to get 19, then square to get 19^2. Let's do the corresponding steps, starting with x: take the square root to get $\sqrt{x}$, add 1 to get $\sqrt{x} + 1$, and then square to get $\boxed{(\sqrt{x} + 1)^2} = x + 2\sqrt{x} + 1$.

Problem 4-5

The probability that a pair of shakers is an identical pair is $q = 1-p$, and all identical pairs are same-colored. The probability that a pair of shakers is a fraternal pair is p. The probability that a random pair is same-colored equals the probability they're identical plus the probability that they're same-color fraternal = $q + \frac{p}{2} = \frac{2q}{2} + \frac{1-q}{2} = \frac{q+1}{2}$. The probability that a same-colored pair is identical = the probability that the pair is identical, which is q, divided by the probability that the pair is same-colored, which is $\frac{q+1}{2}$. The quotient is $\boxed{\dfrac{2q}{q+1}}$.

Problem 4-6

You can get the diagram shown from the original diagram by removing two angle bisectors and using a similar triangle in which the 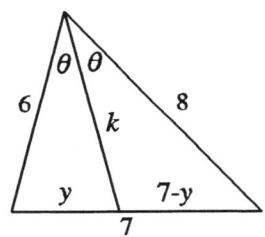 side-lengths have been scaled down by dividing each length by 14. Split the side of length 7 into segments of lengths y and $7-y$. Apply the angle bisector theorem to get $\frac{6}{y} = \frac{8}{7-y}$, from which $y = 3$ and $7-y = 4$. Use the law of cosines twice in the diagram to get $3^2 = 6^2 + k^2 - 12k\cos\theta$ and $4^2 = 8^2 + k^2 - 16k\cos\theta$. Now divide each side of the first of the two equations by 3 and the second by 4, then subtract to eliminate $\cos\theta$. Solve the resulting equation to get $k = 6$. The length of that angle bisector in the unreduced triangle is $6 \times 14 = 84$. That angle bisector marked k splits the side to which it is drawn in the ratio $6{:}8 = 3{:}4$. In the triangle, the length of the right side of the "base" is $(4/7)(98) = 56$. Now, apply the angle bisector theorem to the 84-56-112 triangle. The side with a length of 56 is split into segments whose lengths are in the ratio $84{:}112 = 3{:}4 = 24{:}32$, so $x = \boxed{32}$.

Contests written and compiled by Steven R. Conrad, Daniel Flegler, & Adam Raichel   ©2015 by Mathematics Leagues Inc.

Problem 5-1

Since $x^2+x-2014\times 2015 = x^2-x-2014\times 2015$ simplifies to $x = -x$, the only solution is $x = \boxed{0}$.

Problem 5-2

If $m = \frac{1}{m}$, then $m = 1$ or -1. The equation of line ℓ is either $y = x+1$ or $y = -x-1$. To find the x-intercept of any line, plug in 0 as the y-coordinate. On either line, when $y = 0$, $x = \boxed{-1}$.

Problem 5-3

Using 30°–60°–90° triangle properties in the diagram, we get $AC = \sqrt{3}$, so the perimeter of $\triangle ABC$ is $\boxed{3\sqrt{3}}$.

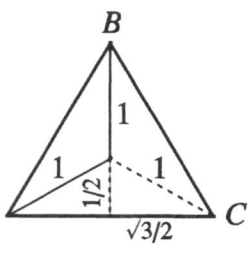

Problem 5-4

If $y = 2^x$, then $y^2 - y - 6 = (y - 3)(y + 2) = 0$. Since 2^x is positive, $y = 2^x = 3$, so $x = \log_2 3$, and $(b,n) = \boxed{(2,3)}$.

Problem 5-5

The perimeter of the rhombus is 60, so each side-length is 15. Also, the length of the diameter of the circle is 25. Let the lengths of the diagonals of the rhombus be $2x$ and $2y$, as shown. By the Pythagorean Theorem, $x^2+y^2 = 15^2$, or $y^2 = 225-x^2$. By the intersecting chords theorem,

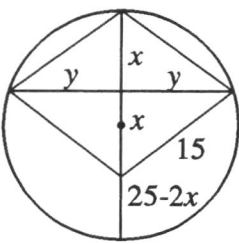

$y^2 = (x)(25-x)$, so $y^2 = 225-x^2 = 25x-x^2$, so $x = 9$, $y = 12$, and the longer diagonal is $\boxed{24}$.

Problem 5-6

Divide the key numbers 1–14 into these 3 groups:

1) numbers divisible by 3 are in set $C = \{3,6,9,12\}$;

2) numbers 1 more than some multiple of 3 are in set $B = \{1,4,7,10,13\}$; and

3) the numbers 2 more than some multiple of 3 are in set $A = \{2,5,8,11,14\}$.

Imagine the ship's steward placing one key in each room, starting in room 1. In room 1, he must leave one of the 5 keys from Group A (leaving 4 remaining in group A). In room 2, he must leave one of the 5 keys from Group B (leaving 4 remaining in group B). In room 3, he must leave one of the 4 keys from Group C (leaving 3 remaining in group C). In room 4, he must leave one of the 4 now-remaining keys from Group A (leaving 3 in group A). The process continues, leaving keys from A, then B, then C, then A, then B, then C, etc., until he has filled all 14 rooms and used all 14 keys. In order, he would have the following number of options for each room from 1 to 14: 5, 5, 4, 4, 4, 3, 3, 3, 2, 2, 2, 1, 1, 1. If we multiply those numbers together, we find that there are $\boxed{345\,600}$ possible ways to place the keys.

Contests written and compiled by Steven R. Conrad, Daniel Flegler, & Adam Raichel   ©2015 by Mathematics Leagues Inc.

Problem 6-1

Each side has an integral length and the length of the longest side is less than half the perimeter. Since the length of the longest side $< 2015/2 = 1007.5$, its greatest possible length is $\boxed{1007}$.

Problem 6-2

Since the value of all Pat's coins is $7.21, the number of pennies Pat has must leave a remainder of 1 when divided by 5. Since Pat has fewer than 100 pennies, Pat has at most $\boxed{96}$ pennies.

Problem 6-3

Part of the original diagram is shown at the right, with each line segment between the centers labeled. We are told that $a+x = 30$, that $b+x = 40$, and $a+x+b = 50$. By subtracting the last equation from the sum of the first two, we get $x = \boxed{20}$.

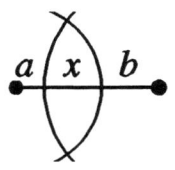

Problem 6-4

$\dfrac{1}{x} + \dfrac{1}{y} = \dfrac{x+y}{xy} = \dfrac{3xy}{xy} = \boxed{3}$.

Problem 6-5

The graphs of $y = f(x)$ and $y = f(-x)$ are reflections across the y-axis, so although the graphs have the same minimum values and maximum values, they have them at opposite x-values. Furthermore, since $f(-x) = \dfrac{x^2+2x+b}{x^2-2x+b} = \dfrac{1}{f(x)}$, the maximum of $f(-x)$ (and thus the maximum of $f(x)$) is the reciprocal of the minimum of $f(x) = $ the reciprocal of $\frac{1}{2} = \boxed{2}$.

[**NOTE:** The maximum is 2 when $b = 9$ and $x = -3$.]

Problem 6-6

The length of any side of a triangle is less than the sum of the lengths of the other two sides. Let's add the known lengths together to obtain an expression we can analyze. Begin with $S = 60\cos A+25\sin A = 5(12\cos A+5\sin A) = (5\times13)\left(\frac{12}{13}\cos A+\frac{5}{13}\sin A\right)$. If $B = \text{Arc}\sin\frac{12}{13} = \text{Arc}\cos\frac{5}{13}$, $S = $ the sum of the known side-lengths $= 65\,(\sin B\cos A+\cos B\sin A) = 65\sin\,(B+A) \le 65$. Since $S \le 65$, the integer-length of the third side $\le 64 < S \le 65$, so the maximum length of the third side is $\boxed{64}$.

Contests written and compiled by Steven R. Conrad, Daniel Flegler, & Adam Raichel   ©2015 by Mathematics Leagues Inc.

Problem 1-1

The sum of the smallest four positive integers is $1+2+3+4 = 10$, and since $10/2 = 5$, the sum we seek is 5. Finally, $1+4 = 5 = 2+3$, so it is possible to get two sets of two integers each whose sum is $\boxed{5}$.

Problem 1-2

We can take at most one 50¢ coin and, with that, at most 1 quarter. Continuing, we can take at most 4 dimes, no nickels, and at most 4 pennies. We can't add more coins without being able to make change for $1, but we can exchange some coins for lower denomination coins. By doing that, we get three sets of coins, $\{1 \times 50¢, 1 \times 25¢, 4 \times 10¢, 4 \times 1¢\}$, $\{3 \times 25¢, 4 \times 10¢, 4 \times 1¢\}$, and $\{1 \times 25¢, 9 \times 10¢, 4 \times 1¢\}$, each with a sum of $\boxed{\$1.19}$.

Problem 1-3

If the expression $|x^2-26x+88| = |x-22| \times |x-4|$ is prime, then one of its factors must be 1. If $|x-22| = 1$, then $x = 23$ or 21. Also, if $|x-4| = 1$, then $x = 5$ or 3. For each of these four values of x, the value of $|x-22| \times |x-4|$ is a prime, so $x = \boxed{3, 5, 21, 23}$.

Problem 1-4

Method I: In the diagram, there are two ways to get the area of the entire shaded triangle. If we use 9 as the base, the altitude is 8. If we use 17 as the base, the altitude is x. Therefore, $9 \times 8 = 17x$, so $x = \boxed{\dfrac{72}{17}}$.

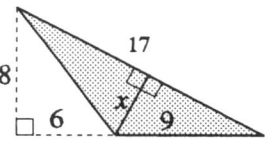

Method II: Using the diagram from Method I, since the rightmost of the two smaller shaded right triangles is similar to the entire 8-15-17 right triangle, we get $9:x = 17:8$. Solving, $x = 72/17$.

Problem 1-5

Any divisor which leaves the same remainder when divided into two different numbers must divide the difference of those two numbers, and so forth. Subtracting, $1453-1108 = 345$, $1844-1453 = 391$, and $2281-1844 = 437$. Continuing, $437-391 = 46$ and $391-345 = 46 = 2(23)$. By inspection, 2 does not work. The required divisor is $\boxed{23}$, which is easily confirmed to leave the same remainder when divided into each of the original numbers.

Problem 1-6

A point with two integral coordinates is called a *lattice point*. Consider the line $5x+7y = c$ for some integer c.

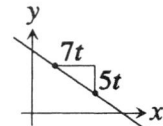

One lattice point that lies on any such line is $(3c,-2c)$. Since the slope of the line is $-5/7$, we can move from any lattice point on the line to another lattice point on the line by moving 7 units to the right and 5 units down (or 7 left, 5 up); and every such line passes through an infinite number of lattice points. If (a,b) is a lattice point on the line, then 5 consecutive lattice points that lie on the line are (a,b), $(a+7,b-5)$, $(a+14,b-10)$, $(a+21,b-15)$, and $(a+28,b-20)$. In fact, every lattice point on the line is of the form $(a+7t,b-5t)$, where t is any integer. The middle 3 points will lie in the first quadrant, and the other 2 will not, if and only if every one of the following conditions is met: $a \le 0$, $a+7 > 0$, $b-20 \le 0$, $b-15 > 0$. We conclude that $-7 < a \le 0$ and $15 < b \le 20$. The largest possible values of a and b (which correspond to the largest possible value of c) that satisfy these inequalities are $a = 0$ and $b = 20$. For $(a,b) = (0,20)$, the 5 consecutive lattice points above are $(0,20)$, $(7,15)$, $(14,10)$, $(21,5)$, and $(28,0)$. We can get the value of c by computing $c = 5x+7y$ for any of these points. This value of c is $\boxed{140}$.

Contests written and compiled by Steven R. Conrad, Daniel Flegler, & Adam Raichel   © 2015 by Mathematics Leagues Inc.

Problem 2-1

Since $23 = 9 + 9 + 4 + 1$, the largest square in this sum is $\boxed{9}$.

[By Lagrange's 4-Square Theorem, every positive integer can be written as the sum of the squares of 4 integers. This is the only way to write 23 as such a sum.]

Problem 2-2

The area of the smaller circle is 16π, so its radius is 4 and a side of the triangle is 8. The larger circle's radius, an altitude of the triangle, is $4\sqrt{3}$. The larger circle's area is $\pi(4\sqrt{3})^2 = \boxed{48\pi}$.

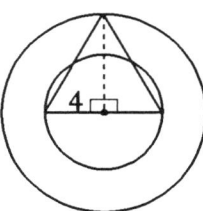

Problem 2-3

We shall show that all ten numbers are equal. If not, then look at the largest one, which is the average of its two nearest neighbors. If one neighbor were smaller, the other would have to be larger. This is impossible, so all three are equal. Similarly, all ten numbers must be equal, so each of the ten numbers is $\boxed{30}$.

Problem 2-4

Clearly, $|x| \neq 1$. To find solutions, consider 2 cases:

Case I: If $|x| > 1$, the exponent must be negative. Since $x^2-x-2 = (x-2)(x+1) < 0 \Leftrightarrow -1 < x < 2$, the solutions in the interval $|x| > 1$ are $\{x \mid 1 < x < 2\}$.

Case II: If $|x| < 1$, the exponent must be positive. Since $x^2-x-2 = (x-2)(x+1) > 0 \Leftrightarrow x < -1$ or $x > 2$, there are no solutions in the interval $|x| < 1$.

Only Case I works, so the solutions are $\boxed{1 < x < 2}$.

Problem 2-5

Method I: Square each side of $n = \sqrt{x\sqrt{x\sqrt{x}}}$ 3 times to get $n^8 = x^7$. If this has a solution in integers, n must be the 7th power of some integer and x must be the 8th power of an integer. The smallest such $x > 1$ must be the 8th power of 2, so $x = 2^8 = \boxed{256}$.

Method II: The innermost x will have its square root taken three times in succession, so the least $x > 1$ for which $\sqrt{x\sqrt{x\sqrt{x}}}$ is an integer must be a power of 2 for which the exponent can be divided by 2 three times in succession. That means that the exponent is $2^3 = 8$, so $x = 2^8 = 256$.

Problem 2-6

Call the integers a, b, and 2015. To minimize $a+b$, make $a < b < 2015$. Since the longest side of a triangle < the sum of the other sides, $\frac{1}{a} < \frac{1}{b} + \frac{1}{2015}$, or $\frac{1}{a} - \frac{1}{b} = \frac{b-a}{ab} < \frac{1}{2015}$. Since $b-a \geq 1$, we get $\frac{1}{ab} \leq \frac{b-a}{ab} < \frac{1}{2015}$, so $ab > 2015$. Since $b > a$, we get $b^2 > 2015$, so $b \geq 45$. If $b = 45$, then $a \leq 44$, so $ab \leq 1980 < 2015$, so this doesn't work. If $b = 46$, then $a \leq 45$. If $a \leq 44$, then $\frac{1}{a} - \frac{1}{b} \geq \frac{1}{44} - \frac{1}{46} = \frac{1}{1012} > \frac{1}{2015}$. If $a = 45$, then $\frac{1}{a} - \frac{1}{b} = \frac{1}{2070} < \frac{1}{2015}$. This works, and $a+b = \boxed{91}$. If $a+b < 91$ and $b \geq 47$, then $a \leq 43$. But then we have $\frac{1}{a} - \frac{1}{b} \geq \frac{4}{2021} > \frac{1}{2015}$, which doesn't work.

Contests written and compiled by Steven R. Conrad, Daniel Flegler, & Adam Raichel   ©2015 by Mathematics Leagues Inc.

Problem 3-1

To maximize the largest angle, minimize the two smallest angles. If the two smallest angles have measures of 1 and 1, the measure of the largest angle of the triangle will be $\boxed{178 \text{ or } 178°}$.

Problem 3-2

The first output is $\frac{1492}{2015}$. This output is fed back in and the second output is 1492 divided by $\frac{1492}{2015}$. That output is $\boxed{2015}$.

Problem 3-3

Method I: By inspection, the sequence is a, a, $2a$, $4a$, $8a$, . . . , so for $n > 2$, each term is twice the preceding term. Therefore, is the 100th term is 2015, the 101st term is $\boxed{4030}$.

Method II: We'll let S_n represent the sum of the first n terms of the sequence, and a_n represent its nth term. We're told that $a_n = S_{n-1}$, so $2015 = a_{100} = S_{99}$, so $a_{101} = a_1 + . . . + a_{99} + a_{100} = S_{99} + a_{100} = 2015 + 2015 = 4030$.

Problem 3-4

In each isosceles triangle, let each vertex angle be $x°$ and let each base angle be $y°$. In each triangle, $x° + 2y° = 180°$. In the diagram, the 12 triangles surround the center, so $10x° + 2y° = 360°$. Subtracting the first equation from the second, $9x° = 180°$, so $x° = \boxed{20°}$.

Problem 3-5

Method I: From the binomial theorem, $(\sqrt{2} - 1)^5 = (\sqrt{2})^5 - 5(\sqrt{2})^4 + 10(\sqrt{2})^3 - 10(\sqrt{2})^2 + 5\sqrt{2} - 1 = 4\sqrt{2} - 5(4) + 10(2\sqrt{2}) - 10(2) + 5\sqrt{2} - 1 = 29\sqrt{2} - 41 = \sqrt{1682} - \sqrt{1681}$, so $k = \boxed{1681}$.

Method II: Use the table feature on your calculator, using $y = \sqrt{x+1} - \sqrt{x}$. Use "ask" for the "independent variable" and "auto" for the "dependent variable." By underestimating and overestimating, you should be able to zero in on $x = 1681$ in not too many tries.

Problem 3-6

Each pair (m,n) has the form (t^2, t^3), where t is a positive integer. Since $m+n = t^2 + t^3 = t^2(t+1)$ is a square, $t+1$ must also be a square. Let $t+1 = k^2$. [Conversely, if $m = t^2$ and $n = t^3$, where $t = k^2 - 1$, then $m+n = t^2 + t^3 = (1+t)(t^2) = k^2(k^2-1)^2 = (k^3-k)^2$ is a square.] Since $m = t^2$, $t = k^2 - 1$, and $m < 1000$, it follows that $(k^2-1)^2 < 1000$. Taking square roots, $k^2 - 1 < 31.6 . . .$, so $k^2 < 32.6 . . . .$ The largest such k^2 is 25, so $t+1 = k^2 = 25$ and $t = 24$. Finally, $m = t^2 = 24^2 = \boxed{576}$.

[Note: $m+n = 24^2 + 24^3 = 576 + 13824 = 14400 = 120^2$.]

Contests written and compiled by Steven R. Conrad, Daniel Flegler, & Adam Raichel   ©2015 by Mathematics Leagues Inc.

Problem 4-1

The statement "$f(x) = 2016$ for all x" means that, for EVERY real number input, the function f assigns the output 2016. Therefore $f(x+2016) = \boxed{2016}$.

Problem 4-2

There are 5 triangles pictured. Let their bases have lengths $b_1, b_2, b_3, b_4,$ and b_5. Let h be the altitude

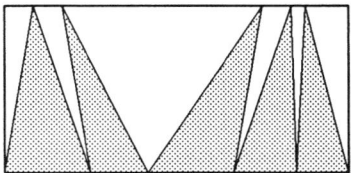

of each triangle [as well as the width of the rectangle]. Hence, the sum of the areas of the 5 triangles is $\frac{1}{2}h(b_1+b_2+b_3+b_4+b_5) = \frac{1}{2}w\ell = \frac{1}{2}(18) = \boxed{9}$.

Problem 4-3

Since $A^{2x} = 4$, $A^x = 2$ and $A^{3x} = 8$. Therefore,

$$\frac{A^{3x} - A^{-3x}}{A^x - A^{-x}} = \frac{8 - \frac{1}{8}}{2 - \frac{1}{2}} = \frac{63/8}{3/2} = \frac{63}{12} = \boxed{\frac{21}{4}}.$$

Problem 4-4

On his 21-minute trip, Al spent 7 minutes running and 14 minutes walking. In km/min, if his walking rate was w, then his running rate was $3w$. Since rate $\times$ time = distance, the distance to work, in km, was $14w + 21w = 35w$. Had he walked for m minutes and run for twice that time, the km distance would have been $wm + (3w)(2m) = 7wm$. Since $35w = 7wm$, $m = 5$. The total time, in minutes, would have been $m + 2m = \boxed{15}$.

Problem 4-5

If we split the 100 integers into 5 disjoint 20-element sets according to the remainder we get when we divide each integer by 5, we get $\{1,6,\ldots,91,96\}$, $\{2,7,\ldots,92,97\}, \{3,8,\ldots,93,98\}, \{4,9,\ldots,94,99\}$, and $\{5,10,\ldots,95,100\}$, where the remainders respectively 1, 2, 3, 4, and 0. To choose as many integers as possible, choose every number from the first and second sets (or first and third sets, or second and fourth sets, or third and fourth sets) and a single number from the fifth set. If we chose any additional number, then, for some number pair, the sum of the remainders would be 0 or 5. We can choose at most $20 + 20 + 1 = \boxed{41}$ of the first 100 positive integers.

Problem 4-6

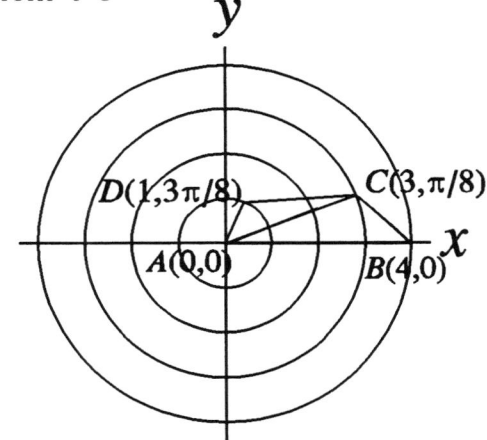

Area of quadrilateral $ABCD$

= area of $\triangle BAC$ + area of $\triangle DAC$

$= \frac{1}{2}AB \times AC \sin \angle BAC + \frac{1}{2}AD \times AC \sin \angle DAC$

$= \frac{1}{2}(4)(3)(\sin\frac{\pi}{8}) + \frac{1}{2}(1)(3)(\sin\frac{2\pi}{8})$

$= 6\sin\frac{\pi}{8} + \frac{3}{2}\sin\frac{\pi}{4}$

$= 3.35676076597\ldots$

$\approx \boxed{3.357}$.

[Note: $6\sin\frac{\pi}{8} + \frac{3}{2}\sin\frac{\pi}{4}$ is an **acceptable** answer.]

Contests written and compiled by Steven R. Conrad, Daniel Flegler, & Adam Raichel   ©2016 by Mathematics Leagues Inc.

Problem 5-1

If $-1 < x < 0$ and n is even, then $x < 0 < x^n$. Similarly, if n is odd, then $-1 < x \le x^n < 0$, with equality if and only if $n = \boxed{1}$.

Problem 5-2

Start at the right and remove one + sign. We get an 89 (no), a 78 (no), a 67 (yes). Removing the sign between 6 and 7, we get $1+2+3+4+5+67+8+9 = 99$, as required. The $\boxed{\text{sixth}}$ + sign must be removed.

[Note: If we remove two plus signs, other solutions are $12+3+4+56+7+8+9$ and $1+23+45+6+7+8+9$.]

Problem 5-3

$\log_b 256 = n \Leftrightarrow b^n = 256$. Since $(2^8)^1 = (2^4)^2 = (2^2)^4 = (2^1)^8 = 256$, it follows that $(b,n) = (2^8,1)$, $(2^4,2)$, $(2^2,4)$, or $(2^1,8)$. The total number of different positive integer values of b is $\boxed{4}$.

Problem 5-4

Use slopes to show that the segments marked as perpendicular are perpendicular. An altitude to the hypotenuse of the heavily outlined right triangle at the bottom of the diagram creates 2 new right triangles similar to each other and to the original triangle. Hence, the smaller shaded right triangle and the lower unshaded right triangle are similar, with ratio of similitude 2:1. If the shorter leg of the smaller of these two triangles is x, that triangle's longer leg (which is the shorter leg of the unshaded right triangle) is $2x$ and the unshaded right triangle's longer leg is $4x$. Each segment that connects a vertex to a midpoint of a non-adjoining side has a length of $x+4x = 5x$, from which the other lengths shown in the diagram are easily obtained. The area of the larger shaded triangle is $150 = (3x)(4x)/2 = 6x^2$, so the area of the smaller shaded right triangle is $(x)(2x)/2 = x^2 = \boxed{25}$.

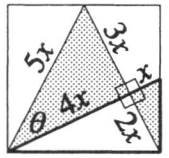

Problem 5-5

In a regular n-gon, each exterior angle is $\frac{360}{n}$ and its adjacent interior angle is $180 - \frac{360}{n}$. Similarly, in a regular m-gon, the measure of each interior angle is $180 - \frac{360}{m}$. The ratio of an interior angle of M to an interior angle of N is $\frac{3}{2}$. Set $\frac{3}{2}$ equal to the complex fraction created when we form the ratio and solve. We get $360 - \frac{720}{m} = 540 - \frac{1080}{n}$. This simplifies to $\frac{6}{n} - \frac{4}{m} = 1$. It's clear that $n < 6$. Trying $n = 3$, 4, and 5, we get $(m,n) = \boxed{(4,3),\ (8,4),\ (20,5)}$.

Problem 5-6

If $b = \#$ of bronze medals in the box, we have that $P(\text{gold medal}) = \frac{20}{30+b}$, $P(\text{silver medal}) = \frac{10}{30+b}$, and $P(\text{bronze medal}) = \frac{b}{30+b}$. Consequently, $P(2$ gold medals and 2 silver medals$) = {_4}C_2 \times \left(\frac{20}{30+b}\right)^2 \times \left(\frac{10}{30+b}\right)^2$, and $P(1$ gold, 1 silver, 2 bronze medals$) = {_2}C_1 \times \left(\frac{20}{30+b}\right) \times \left(\frac{10}{30+b}\right) \times {_4}C_2 \times \left(\frac{b}{30+b}\right)^2$. If we equate these last two probabilities and solve, we get $200 = 2b^2$, so $b = \boxed{10}$.

Contests written and compiled by Steven R. Conrad, Daniel Flegler, & Adam Raichel   ©2016 by Mathematics Leagues Inc.

Problem 6-1

The rectangles are not congruent, so the answer is the least positive integer that can be factored in more than one way. Since $4 = 1 \times 4 = 2 \times 2$, the area of each rectangle is $\boxed{4}$.

Problem 6-2

By the Pythagorean Theorem, the square of the hypotenuse equals $\left(\frac{\pi}{3}\right)^2 + \left(\frac{\pi}{4}\right)^2 = \frac{25\pi^2}{144}$. To get the length of the hypotenuse, take the square root of the result. That equals $\boxed{\frac{5\pi}{12}}$.

Problem 6-3

If we call the common solution x_1, then it follows that $x_1^3 + kx_1^2 - 3x_1 + 4 = x_1^3 + kx_1^2 - 5x_1 + 8$, from which we subtract to see that $x_1 = 2$. Substitute back into the first of the original equations to get $2^3 + k(2^2) - 3(2) + 4 = 0$, from which $k = \boxed{\frac{-3}{2}}$.

Problem 6-4

Method I: Draw vertical line segments connecting each vertex of the polygon to the vertex directly above or below it, thereby creating 10 rhombuses, each with sides of length 1. Dropping an altitude of one of the rhombuses creates an isosceles right triangle whose legs each have length $\frac{\sqrt{2}}{2}$. Finally, the polygon's area is equal to $10bh = 10(1)\left(\frac{\sqrt{2}}{2}\right) = \boxed{5\sqrt{2}}$.

Method II: The total area of all ten rhombuses is equal to $10 \times (1)(1)(\sin 45°) = 5\sqrt{2}$.

Problem 6-5

Since $\sin 2016° = -\sin 36° = -\cos 54°$, it follows that $\sin^2 2016° = \cos^2 54°$. Rewrite the original equation as $\cos^2 54° + \sin^2 x = 1$, which has 2 sets of solutions. In one set, with n an integer, we have $x = 54 + 180n$, $0 \le n \le 10$. There are 11 solutions in this set. In the other set, also with n an integer, we have $x = 126 + 180n$, $0 \le n \le 10$. There are also 11 solutions in this set. Altogether, the total number of solutions is $11 + 11 = \boxed{22}$.

Problem 6-6

Let p = the probability that the first person wins. The probability that the first person wins by rolling a 9 on the first roll is $\frac{4}{36} = \frac{1}{9}$. The probability that he does not roll a 9 on his first turn is $\frac{8}{9}$. If the first person does not roll a 9 on his first turn, then the second person, at his first turn, will have probability p of winning, since he will then have the same chance of winning as the first person had on his first turn. The probability that the second person wins is therefore the product of the probability that the first person does not roll a 9 times the probability that the second person wins when the first person does not roll a 9 on his first roll = $\frac{8}{9}p$. Solving $p + \frac{8}{9}p = 1$, we get $p = \frac{9}{17}$. The ratio of the probability that the first person wins to the probability that the second wins is 9:8 = 450:400, so the number of dollars that the first person should pay is $\boxed{450 \text{ or } \$450}$.

Contests written and compiled by Steven R. Conrad, Daniel Flegler, & Adam Raichel   ©2016 by Mathematics Leagues Inc.

Answers &
Difficulty Ratings

October, 2011 — March, 2016

Answers

	2011-2012		**2012-2013**		**2013-2014**
1-1.	(1,1)	1-1.	$\frac{\pi}{2}$	1-1.	18
1-2.	1	1-2.	2, 2012	1-2.	168
1-3.	X	1-3.	49	1-3.	(1,4,8)
1-4.	60	1-4.	7	1-4.	18 112
1-5.	2	1-5.	502	1-5.	$\sqrt{20}$
1-6.	1 000 005	1-6.	24	1-6.	9%
2-1.	0	2-1.	37	2-1.	11-12-13
2-2.	$\frac{1}{4}$ or 0.25	2-2.	12	2-2.	−1
2-3.	2011	2-3.	8	2-3.	17
2-4.	$2 + 2\sqrt{3}$	2-4.	(3,3,3), (2,4,4), (2,3,6)	2-4.	206
2-5.	42 857	2-5.	100	2-5.	98
2-6.	5	2-6.	4026	2-6.	108
3-1.	32	3-1.	20	3-1.	12
3-2.	1, 3, 5, 8, 8	3-2.	4024	3-2.	54
3-3.	$\frac{1}{3}$	3-3.	(20,19)	3-3.	4 050 157
3-4.	27	3-4.	214	3-4.	0
3-5.	4022	3-5.	$\left(\frac{7}{2},\frac{11}{2}\right)$	3-5.	2013
3-6.	$\frac{3\sqrt{2}}{4}$	3-6.	$\frac{7}{72}$	3-6.	−1, 2, 5, 6, 9
4-1.	2012^2	4-1.	−2013	4-1.	12
4-2.	$\frac{1}{29}, \frac{1}{30}$	4-2.	48	4-2.	1
4-3.	February	4-3.	25	4-3.	2.5
4-4.	45	4-4.	7	4-4.	$\sqrt{3}$
4-5.	$\frac{4}{5}$	4-5.	106π	4-5.	314
4-6.	(7,4)	4-6.	720	4-6.	$\frac{7}{16}$
5-1.	$2012\sqrt{\pi}$	5-1.	12	5-1.	1
5-2.	49	5-2.	50	5-2.	0
5-3.	$\frac{1 \pm \sqrt{5}}{2}$	5-3.	968	5-3.	6
5-4.	40, 60	5-4.	$\frac{1}{2013}$	5-4.	62
5-5.	(1,2,3)	5-5.	45	5-5.	2, 2i, −2, −2i
5-6.	$\sqrt[3]{\frac{1}{2}}$	5-6.	42, 46	5-6.	$\frac{a^3 + 3a^2}{2}$
6-1.	(2011,2012)	6-1.	4028	6-1.	2,−1
6-2.	31	6-2.	(25,23)	6-2.	14
6-3.	(2,3,5)	6-3.	4	6-3.	72
6-4.	6	6-4.	5	6-4.	4/3,12
6-5.	$-\frac{1}{2}$	6-5.	(13,13)	6-5.	−1/8
6-6.	10	6-6.	130π	6-6.	$\sqrt[8]{2014}$

Answers

<table>
<tr><td colspan="2">**2014-2015**</td><td colspan="2">**2015-2016**</td></tr>
<tr><td>1-1.</td><td>1</td><td>1-1.</td><td>5</td></tr>
<tr><td>1-2.</td><td>465 209</td><td>1-2.</td><td>$1.19</td></tr>
<tr><td>1-3.</td><td>2012</td><td>1-3.</td><td>3, 5, 21, 23</td></tr>
<tr><td>1-4.</td><td>180 or 180°</td><td>1-4.</td><td>$\frac{72}{17}$</td></tr>
<tr><td>1-5.</td><td>(3,24)</td><td>1-5.</td><td>23</td></tr>
<tr><td>1-6.</td><td>*AAKAAKKK*</td><td>1-6.</td><td>140</td></tr>
<tr><td></td><td></td><td></td><td></td></tr>
<tr><td>2-1.</td><td>4π</td><td>2-1.</td><td>9</td></tr>
<tr><td>2-2.</td><td>2014</td><td>2-2.</td><td>48π</td></tr>
<tr><td>2-3.</td><td>43</td><td>2-3.</td><td>30</td></tr>
<tr><td>2-4.</td><td>4</td><td>2-4.</td><td>$1 < x < 2$</td></tr>
<tr><td>2-5.</td><td>62</td><td>2-5.</td><td>256</td></tr>
<tr><td>2-6.</td><td>5005</td><td>2-6</td><td>91</td></tr>
<tr><td></td><td></td><td></td><td></td></tr>
<tr><td>3-1.</td><td>4028</td><td>3-1.</td><td>178 or 178°</td></tr>
<tr><td>3-2.</td><td>500</td><td>3-2.</td><td>2015</td></tr>
<tr><td>3-3.</td><td>270</td><td>3-3.</td><td>4030</td></tr>
<tr><td>3-4.</td><td>0</td><td>3-4.</td><td>20°</td></tr>
<tr><td>3-5.</td><td>$3 < x < 12$</td><td>3-5.</td><td>1681</td></tr>
<tr><td>3-6.</td><td>53</td><td>3-6.</td><td>576</td></tr>
<tr><td></td><td></td><td></td><td></td></tr>
<tr><td>4-1.</td><td>2015</td><td>4-1.</td><td>2016</td></tr>
<tr><td>4-2.</td><td>4</td><td>4-2.</td><td>9</td></tr>
<tr><td>4-3.</td><td>12</td><td>4-3.</td><td>$\frac{21}{4}$</td></tr>
<tr><td>4-4.</td><td>$x + 2\sqrt{x} + 1$</td><td>4-4</td><td>15</td></tr>
<tr><td>4-5.</td><td>$\frac{2q}{q+1}$</td><td>4-5.</td><td>41</td></tr>
<tr><td>4-6.</td><td>32</td><td>4-6.</td><td>$6\sin\frac{\pi}{8} + \frac{3}{2}\sin\frac{\pi}{4} \approx 3.357$</td></tr>
<tr><td></td><td></td><td></td><td></td></tr>
<tr><td>5-1.</td><td>0</td><td>5-1.</td><td>1</td></tr>
<tr><td>5-2.</td><td>−1</td><td>5-2.</td><td>sixth</td></tr>
<tr><td>5-3.</td><td>$3\sqrt{3}$</td><td>5-3.</td><td>4</td></tr>
<tr><td>5-4.</td><td>(2,3)</td><td>5-4.</td><td>25</td></tr>
<tr><td>5-5.</td><td>24</td><td>5-5.</td><td>(4,3), (8,4), (20,5)</td></tr>
<tr><td>5-6.</td><td>345 600</td><td>5-6.</td><td>10</td></tr>
<tr><td></td><td></td><td></td><td></td></tr>
<tr><td>6-1.</td><td>1007</td><td>6-1.</td><td>4</td></tr>
<tr><td>6-2.</td><td>96</td><td>6-2.</td><td>$\frac{5\pi}{12}$</td></tr>
<tr><td>6-3.</td><td>20</td><td>6-3.</td><td>$\frac{-3}{2}$</td></tr>
<tr><td>6-4.</td><td>3</td><td>6-4.</td><td>$5\sqrt{2}$</td></tr>
<tr><td>6-5.</td><td>2</td><td>6-5.</td><td>22</td></tr>
<tr><td>6-6.</td><td>64</td><td>6-6.</td><td>450 or $450</td></tr>
</table>

Difficulty Ratings

(% correct of all reported scores from each participating school)

2011-2012		2012-2013		2013-2014		2014-2015		2015-2016	
1-1.	64%	1-1.	75%	1-1.	87%	1-1.	80%	1-1.	78%
1-2.	85%	1-2.	57%	1-2.	76%	1-2.	54%	1-2.	30%
1-3.	57%	1-3.	37%	1-3.	68%	1-3.	42%	1-3.	30%
1-4.	46%	1-4.	44%	1-4.	48%	1-4.	51%	1-4.	43%
1-5.	22%	1-5.	9%	1-5.	14%	1-5.	53%	1-5.	36%
1-6.	24%	1-6.	11%	1-6.	11%	1-6.	45%	1-6.	11%
2-1.	85%	2-1.	50%	2-1.	82%	2-1.	92%	2-1.	83%
2-2.	49%	2-2.	93%	2-2.	77%	2-2.	62%	2-2.	69%
2-3.	81%	2-3.	62%	2-3.	41%	2-3.	43%	2-3.	58%
2-4.	31%	2-4.	47%	2-4.	21%	2-4.	50%	2-4.	36%
2-5.	43%	2-5.	27%	2-5.	29%	2-5.	68%	2-5.	62%
2-6.	16%	2-6.	9%	2-6.	13%	2-6.	3%	2-6.	7%
3-1.	75%	3-1.	71%	3-1.	86%	3-1.	90%	3-1.	87%
3-2.	86%	3-2.	70%	3-2.	89%	3-2.	52%	3-2.	81%
3-3.	69%	3-3.	75%	3-3.	41%	3-3.	67%	3-3.	75%
3-4.	41%	3-4.	33%	3-4.	20%	3-4.	36%	3-4.	79%
3-5.	34%	3-5.	19%	3-5.	26%	3-5.	13%	3-5.	55%
3-6.	4%	3-6.	4%	3-6.	28%	3-6.	2%	3-6.	26%
4-1.	81%	4-1.	84%	4-1.	92%	4-1.	86%	4-1.	66%
4-2.	73%	4-2.	76%	4-2.	49%	4-2.	48%	4-2.	84%
4-3.	50%	4-3.	72%	4-3.	40%	4-3.	37%	4-3.	49%
4-4.	64%	4-4.	27%	4-4.	23%	4-4.	62%	4-4.	54%
4-5.	20%	4-5.	11%	4-5.	14%	4-5.	19%	4-5.	29%
4-6.	11%	4-6.	14%	4-6.	13%	4-6.	21%	4-6.	10%
5-1.	67%	5-1.	67%	5-1.	92%	5-1.	92%	5-1.	53%
5-2.	55%	5-2.	89%	5-2.	62%	5-2.	76%	5-2.	82%
5-3.	29%	5-3.	79%	5-3.	32%	5-3.	59%	5-3.	49%
5-4.	38%	5-4.	17%	5-4.	52%	5-4.	44%	5-4.	25%
5-5.	36%	5-5.	12%	5-5.	20%	5-5.	30%	5-5.	18%
5-6.	6%	5-6.	18%	5-6.	4%	5-6.	20%	5-6.	16%
6-1.	86%	6-1.	74%	6-1.	88%	6-1.	69%	6-1.	65%
6-2.	79%	6-2.	70%	6-2.	85%	6-2.	85%	6-2.	82%
6-3.	67%	6-3.	69%	6-3.	46%	6-3.	88%	6-3.	44%
6-4.	23%	6-4.	31%	6-4.	55%	6-4.	67%	6-4.	47%
6-5.	42%	6-5.	33%	6-5.	68%	6-5.	45%	6-5.	25%
6-6.	65%	6-6.	8%	6-6.	33%	6-6.	17%	6-6.	17%

Math League Contest Books

4th Grade Through High School Levels

Written by Steven R. Conrad and Daniel Flegler, recipients of President Reagan's 1985 Presidential Awards for Excellence in Mathematics Teaching, each book provides you with problems from *regional* mathematics competitions.

Order books at www.mathleague.com (or use the form below)

Name _____

Address _____

City _____ State _____ Zip _____
                                     (*or Province*)   (*or Postal Code*)

Available Titles	**# of Copies**	**Cost**
Math Contests—Grades 4, 5, 6	($12.95 per book)	
Volume 1: 1979-80 through 1985-86	_____	_____
Volume 2: 1986-87 through 1990-91	_____	_____
Volume 3: 1991-92 through 1995-96	_____	_____
Volume 4: 1996-97 through 2000-01	_____	_____
Volume 5: 2001-02 through 2005-06	_____	_____
Volume 6: 2006-07 through 2010-11	_____	_____
Volume 7: 2011-12 through 2015-16	_____	_____
Math Contests—Grades 7 & 8‡	‡(Vols. 3, 4, 5, 6, & 7 include Algebra I)	
Volume 1: 1977-78 through 1981-82	_____	_____
Volume 2: 1982-83 through 1990-91	_____	_____
Volume 3: 1991-92 through 1995-96	_____	_____
Volume 4: 1996-97 through 2000-01	_____	_____
Volume 5: 2001-02 through 2005-06	_____	_____
Volume 6: 2006-07 through 2010-11	_____	_____
Volume 7: 2011-12 through 2015-16	_____	_____
Math Contests—High School		
Volume 1: 1977-78 through 1981-82	_____	_____
Volume 2: 1982-83 through 1990-91	_____	_____
Volume 3: 1991-92 through 1995-96	_____	_____
Volume 4: 1996-97 through 2000-01	_____	_____
Volume 5: 2001-02 through 2005-06	_____	_____
Volume 6: 2006-07 through 2010-11	_____	_____
Volume 7: 2011-12 through 2015-16	_____	_____
Shipping and Handling	$3 ($5 Canadian)	

Please allow 4-6 weeks for delivery

Total: $_____

☐ Check or Purchase Order Enclosed; **or**

☐ Visa / MasterCard / Discover # _____

☐ Expiration Date _____ Signature _____

Mail your order with payment to:

Math League Press, P.O. Box 17, Tenafly, NJ 07670-0017
or order on the Web at www.mathleague.com

Phone: (201) 568-6328 • Fax: (201) 816-0125